Terahertz
Liquid
Photonics

Other Titles by Xi-Cheng Zhang

Advanced Semiconductor Physics (in Chinese)
THz Science, Technology, and Applications (in Chinese)
Introduction to THz Wave Photonics

Terahertz Liquid Photonics

Edited by

Xi-Cheng Zhang
University of Rochester, USA

Yiwen E
University of Rochester, USA

Liangliang Zhang
Capital Normal University, China

Anton Tcypkin
ITMO University, Russia

W⊖ World Scientific

NEW JERSEY · LONDON · SINGAPORE · BEIJING · SHANGHAI · HONG KONG · TAIPEI · CHENNAI · TOKYO

Published by

World Scientific Publishing Co. Pte. Ltd.

5 Toh Tuck Link, Singapore 596224

USA office: 27 Warren Street, Suite 401-402, Hackensack, NJ 07601

UK office: 57 Shelton Street, Covent Garden, London WC2H 9HE

Library of Congress Cataloging-in-Publication Data
Names: Zhang, Xi-Cheng, 1956–   editor. | E, Yiwen, editor. | Zhang, Liangliang, editor. |
    Tcypkin, Anton, editor.
Title: Terahertz liquid photonics / edited by Xi-Cheng Zhang, Yiwen E,
    Liangliang Zhang, Anton Tcypkin.
Description: New Jersey : World Scientific, [2024] | Includes bibliographical references.
Identifiers: LCCN 2023004292 | ISBN 9789811265631 (hardcover) |
    ISBN 9789811265648 (ebook for institutions) | ISBN 9789811265655 (ebook for individuals)
Subjects: LCSH: Terahertz technology. | Photonics.
Classification: LCC TK7877 .T439 2024 | DDC 621.36/5--dc23/eng/20230531
LC record available at https://lccn.loc.gov/2023004292

British Library Cataloguing-in-Publication Data
A catalogue record for this book is available from the British Library.

For any available supplementary material, please visit
https://www.worldscientific.com/worldscibooks/10.1142/13122#t=suppl

Desk Editor: Shaun Tan Yi Jie

Typeset by Stallion Press
Email: enquiries@stallionpress.com

Preface

The far-infrared region (0.3–10 THz) has long been considered the last remaining scientific gap in the electromagnetic spectrum, underdeveloped but ripe for exploitation. As a bridge between electronics and photonics, terahertz (THz) science and technology have made tremendous progress in the past decades. To provide broadband, high-field THz pulses for fundamental research and technological application, generations of THz pulses with femtosecond laser excitation have been widely studied. Three of the four states — solids, gases, and plasmas — have been used as the sources for emitting THz waves for decades. However, the use of liquids as THz wave emitters has been historically sworn off. One major reason is the high absorption shown in polar liquids at THz range, which may hinder the development of THz liquid photonics. Compared to other matter states, liquids present numerous unique properties. Specifically, liquids have a comparable material density to that of solids, meaning that laser pulses over a certain area will interact with three orders more molecules than an equivalent cross-section of gas. In contrast with solids, the fluidity of liquids allows each laser pulse to interact with a fresh area of the target. Therefore, material damage threshold is not an issue even with laser pulses of high repetition rate. This makes liquids very promising candidates for the study of high energy density plasma, and ultrafast dynamics of ionized particles in the process of laser-matter interaction.

THz liquid photonics is a newly emerging topic in recent years, offering an alternative option for researchers to obtain THz emission. This

interdisciplinary and transformative topic should lead to key progress that will enable new science and advance numerous THz wave sensing and spectroscopy technologies, including next-generation liquid source, device, and system development. Since the first report of THz wave emission from liquids under short laser pulse excitation in 2017, numerous investigations in this field have been conducted and published from researchers worldwide in a short period. "THz liquid photonics" as a new topic has been listed in several international conferences.

The authors of this book are from the most active researchers in the THz liquid photonics community. The book includes the recent experimental results, theoretical analysis, and simulated calculations on THz emission and detection from liquid materials under ultrashort-pulse laser excitation, providing readers with a comprehensive understanding of current developments in the field. By comparing with traditional sources, distinctive properties of THz wave generation from liquids are discussed in detail, which provides a new perspective in exploring laser-liquid interactions. THz liquid photonics will be a new frontier in the THz community.

Our authors would like to thank Qi Jin, Kareem Garriga Francis, Fang Ling, Yuqi Cao, and Steven Fu. Qi Jin was the first student who successfully demonstrated THz wave generation from flowing water in the Institute of Optics at the University of Rochester in 2017, which opened a new area on THz wave liquid photonics. There were several students who worked on this project for more than 10 years at Rensselaer Polytechnic Institute and University of Rochester, but Qi Jin was the only one who believed it and continued the experiments. Prof. Jianming Dai provided essential technical support during the experiments.

Xi-Cheng Zhang (University of Rochester, USA)
Yiwen E (University of Rochester, USA)
Liangliang Zhang (Capital Normal University, China)
Anton Tcypkin (ITMO University, Russia)
August 2022

About the Editors

Dr. Xi-Cheng Zhang is Endowed Parker Givens Chair Professor of Optics at the University of Rochester, USA, where he is also Professor of Physics and Distinguished Scientist at the Laboratory for Laser Energetics. He is Honorary Professor at Jilin University, China and Lomonosov Moscow State University, Russia. He is an Elected Fellow of the American Physical Society (APS), American Association for the Advancement of Science (AAAS), Institute of Electrical and Electronics Engineers (IEEE), Society of Photo-Optical Instrumentation Engineers (SPIE) and Optical Society of America (OSA), as well as a Foreign Member of the Russian Academy of Sciences. He is co-Editor-in-Chief for *Light, Science, and Applications*, ranked 3rd out of 99 journals in the optics category.

Professor Zhang's research interests center around terahertz waves, in particular the generation, detection, and applications of free-space THz beams with ultrafast optics. He has published >350 peer-reviewed papers, delivered >700 talks, owns 29 patents and wrote 28 books and book chapters, with a h-index of 94 (June 2022). He has received many awards and honors over the years, including the Alexander von Humboldt Foundation Prize (2018), the International Society of Infrared, Millimeter, and Terahertz Waves Kenneth F. Button Prize (2014), the OSA William F.

Meggers Award (2012), the IEEE Photonics Society William Streifer Scientific Achievement Award (2011), and Rensselaer's William H. Wiley Award (2009).

Dr. Yiwen E is Research Associate at the University of Rochester, USA. Her research experience includes THz wave generation from ionized gas/liquid/solid, EM waves interaction with metamaterials/biomaterials, phonon-polariton regulated propagation of THz waves in crystals, and THz time domain spectroscopy systems. She received the Institute Director Scholarship (2012–2015) and Excellent Graduate Award (2012–2014) from the University of Chinese Academy of Sciences, China, where she holds a PhD in Optics. She was the winner of the second-class prize of the 1st Women in Ultrafast Science Global Award in 2022.

Dr. Liangliang Zhang is Professor at the Capital Normal University, China. Her main research interests are intense THz sources, THz aqueous & air photonics, as well as THz spectroscopy and imaging technologies. She has published more than 80 papers in leading academic journals such as *Physical Review Letters*, *Light: Science and Applications*, *Applied Physics Letters*, and *Optics Letters*, and owns 12 Chinese invention patents. She received the First Prize of Scientific and Technological Progress of China Instrument and Control Society (2021), First Prize of Scientific Research Excellence Award (Science and Technology) of the Chinese Ministry of Education (2013), National Excellent Doctoral Dissertation of China (2011), Excellent Doctoral Dissertation of Beijing (2010), and Wang Daheng Prize of the Chinese Optical Society (2007).

Dr. Anton Tcypkin is Director, Associate Professor at the Faculty of Photonics and Optoinformatics, the head of the International Laboratory of Femtosecond Optics and Femtotechnology, and the head of the Laboratory of Quantum Process and Measurements, ITMO University, Russia. He was the Chairman of the organizing committee of the international conference *Fundamental Problems of Optics* (2017–2020). He has published more than 100 papers, delivered more than 70 talks, and owns 5 patents. He received the Medal V S Letohova (2017), Scholarship of the President of the Russian Federation (2015–2017), and the title of Best Student of ITMO University (2009). He is an Expert of the Russian Science Foundation (2016–), and a member of the National Experts Collegium of CIS countries on lasers and laser technologies (2022–).

Contents

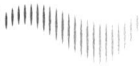

Introduction

Yiwen E, X.-C. Zhang

The Institute of Optics, University of Rochester, USA

An understanding of the fundamentals in laser-liquid interaction and emission of electromagnetic waves from laser-ionized materials is crucial for the development of new techniques in optics, biology, medical diagnosis, and therapy. With tremendous focus on the ultrashort pulsed solid-state laser, laser intensity has been improved at breakneck speed in recent years. From the threshold of plasma formation at approximately 10^{10} W/cm² to the current world record intensity of 10^{23} W/cm², the dynamics of electrons or charged particles have entered the relativistic region. Many interesting processes including field-induced current, nonlinear propagation of laser pulses, coherent transition radiation, sheath field acceleration, etc., have been considerably studied.

From a macroscopic point of view, several nonlinear optical effects or interactions are involved when a strong laser is applied to a medium, including supercontinuum generation, optical self-focusing, plasma defocusing, optical intensity camping, and so on. With ionization, the dynamics of electrons highly depend on the interaction between the evolved optical field and the medium. From a microscopic point of view, when an

atom or a molecule is ionized by an intense laser field, the free electron will initiate a series of reactions. In an aqueous system, the radiolysis of water plays a vital role in the process of water disinfection, corrosion damage in nuclear waste storage, interfacial chemistry, radiotherapy, etc. Despite intensive investigations into the radiolysis of liquid media, a lack of comprehensive understanding of this process remains.

Terahertz (THz) wave is in the electromagnetic spectrum with its frequency range from 0.1 to 10 THz, located between microwave and far-infrared wave. This narrow gap makes the THz wave the highest frequency for electronics, and the lowest frequency for photonics. In other words, THz waves bridge traditional electronics and photonics, and pave the way for development of a photo-electron hybrid system. With efforts from past decades, THz technology has started playing an important role in scientific research. It has helped us study molecular rotational and vibrational transitions, crystal vibrations, and hydrogen bond vibrations. With the development of THz pulses with a field strength greater than MV/cm, THz field-induced nonlinear effect has become a necessary "tool" for researchers to investigate strong field-induced non-resonant modulation. We believe THz liquid photonics will provide a new perspective in understanding the laser-induced ionization of liquid media.

In this book, we review the state-of-art in laser-liquid systems and raise critical research issues and fundamental questions mainly based on the recent development of THz wave liquid photonics. In particular, we summarize the observation of broadband THz wave generation from liquids with different geometries (film, line, droplet, etc.). For detection, THz field-induced nonlinear effects in liquids are discussed. Based on these observations, this book provides an understanding of the laser-induced ionization process in liquids for both microscopic and macroscopic views.

© 2024 World Scientific Publishing Company
https://doi.org/10.1142/9789811265648_0001

Chapter 1
Terahertz Wave Generation via Laser-Produced Ionization

Yiwen E, X.-C. Zhang

The Institute of Optics, University of Rochester, USA

1.1 Introduction

Laser-induced ionization happens in media when electrons in bound states absorb enough photon energy to overcome the work function and transfer to excited states (gas, molecule) or continua (liquid, solid) [1–4]. These excited electrons are free to move in a certain area to create transient currents radiating electromagnetic waves. Direct photon ionization happens when the single photon energy is equal to or greater than the ionization potential. If lower, there is a probability that the ionization happens via multiphoton or tunnel ionization when the laser intensity is high enough. Fig. 1.1 shows the schematic diagram illustrating multiphoton and tunnel ionizations. In a multiphoton ionization process, as shown in Fig. 1.1(a), the electron absorbs several photons simultaneously to escape from the bound state. For tunnel ionization as shown in Fig. 1.1(b), the potential barrier is distorted with the existence of an external electric field, resulting in a higher probability for the electrons to tunnel through the potential barrier. Usually, the Keldysh parameter [5] is used to distinguish these two processes, and can

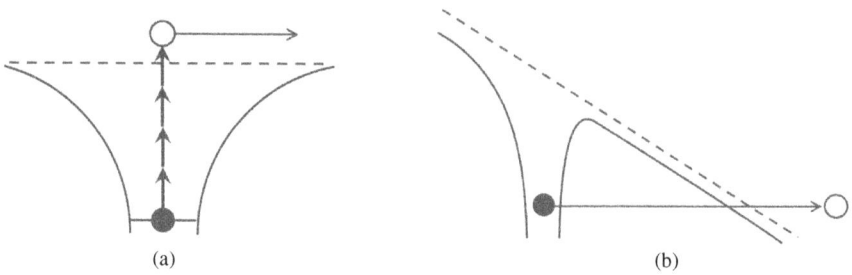

▲ Fig. 1.1.   Schematic diagram for (a) multiphoton ionization and (b) tunnel ionization.

be calculated by the frequency and amplitude of the driving field, as well as the ionization potential.

Terahertz (THz) wave generation from laser-induced ionization has emerged as one of the most promising solutions for obtaining broadband, high-intensity THz pulses. Hamster *et al.* first reported THz wave emission from ionized material in 1993 [6], in which both compressed gas and an aluminum-coated glass were used as targets. This observation has attracted much attention because it allows researchers to fully use the laser pulse energy produced by the modern chirped pulse amplification without worrying about the damage of targets/emitters. Accordingly, air plasma has become one of the most popular THz sources due to its simplicity in implementation, in which only one single lens is needed to focus intense laser pulses into the air. An important milestone in the development of THz emission from ionized gas material is the significant enhancement in generation efficiency by using the two-color excitation geometry, which was reported by Cook and Hochstrasser in 2000 [7]. In this geometry, both the fundamental beam and its second harmonic beam generated by a nonlinear crystal are focused onto a target. With a much higher generation efficiency, a THz peak field over MV/cm has been observed from a two-color air plasma source, making ionized material a promising source for the study of THz field-induced nonlinear effects [8]. Fig. 1.2 showcases the schematic diagram of THz emission from (a) single-color and (b) two-color air plasma. For a two-color geometry, a frequency-doubling crystal is used to generate the second harmonic wave. The THz field can be coherently

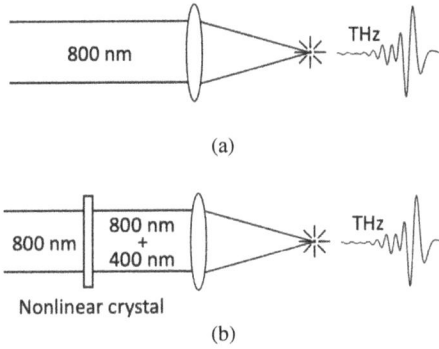

(a)

(b)

▲ Fig. 1.2.   Schematic diagram for THz emission from (a) single-color laser-induced air plasma and (b) two-color laser-induced air plasma.

controlled by finely tuning the relative phase between the fundamental and the second harmonic waves [9–14]. It should be noted that the radiation patterns of THz waves from these two cases are different.

Moreover, THz emission from ionized solid material exhibits advantages in generating intense THz waves and investigating laser-matter interaction in relativistic regimes. In 2008, THz wave generation from the surface of a solid target was reported by Sagisaka *et al.* [15], in which electrons ejected under laser excitation generate a transient current on the target surface. Since then, the record of THz pulse energy has been constantly reported from solid targets with a single-shot geometry. Recently, it has been observed by Liao *et al.* that THz pulse energy exceeding millijoule (mJ) is generated from picosecond laser-irradiated metal foils [16]. In this chapter, several mechanics of THz emission from ionized material are discussed in detail.

1.2 Ponderomotive Force-Induced Transient Current

In the process of THz emission under single-color optical excitation, there are two THz radiation mechanisms. One is ponderomotive force-induced current and another is the second-order nonlinear effect in the spatially inhomogeneous plasma. For the single-color air plasma, the

ponderomotive force [17] from the inhomogeneous electric field at the focus dominates the contribution of THz emission. The ponderomotive force is expressed as

$$\vec{F} = -\frac{q^2}{2nm\,\omega^2}\nabla(\vec{E}\cdot\vec{E}) \tag{1.1}$$

where q is the electrical charge of the particle, n is the charge density, m is the particle mass, ω is the laser angular frequency and E is the laser electric field. Since the electron mass is much smaller than other particles, they are accelerated at a much higher speed in a short time. We will focus on the movement of electrons, which mainly contribute to the emission of THz waves. Based on Eq. (1.1), ionized electrons under an inhomogeneous field are pushed toward the area with lower field intensity [18]. Assuming that the laser beam propagates in $+z$ direction, plasma density remains identical in the laser propagation direction since electrons cannot move faster than the laser beam. Therefore, the overall acceleration direction of electrons is in the $-z$ direction, creating a dipole along the laser propagation direction. When the laser propagates away from the plasma, the restoring force brings electrons back to ions. This process induces a transient current oscillation with a period close to the time duration of the optical pulse, radiating electromagnetic waves in the THz range with a conical energy distribution [19]. The conical angle highly depends on the geometry length of the plasma. A longer filament length results in a smaller conical angle. Fig. 1.3 shows a fluorescence image of an air plasma from the sideway. The arrow shows the laser propagation direction. A two-inch focal length lens is used in this case. The central wavelength of the laser pulse is 800 nm.

The longitudinal length of a laser-induced plasma is usually longer than the Rayleigh length of the laser beam because the plasma starts where the intensity is over the ionization threshold, and then experiences multiple cycles of self-focusing and defocusing [20, 21]. When a high numerical-aperture (NA) lens is used for focusing, it is possible to create a plasma with significantly low laser energies. "Microplasma" refers to plasma with a longitudinal length smaller than 1 mm [22, 23]. The minimum energy

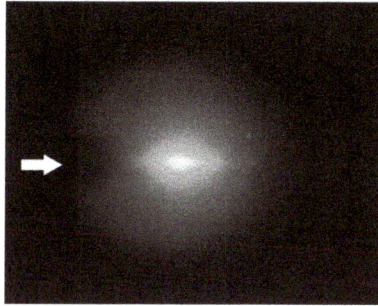

▲ Fig. 1.3.   Image of a single-color air plasma taken from the side with a commercial CCD camera. The arrow indicates the laser propagation direction. Reprinted with permission from Zhu L.-G., Sheng Z., Schneider H., Chen H.-T. & Tani M. (2022). Ultrafast phenomena and terahertz waves: introduction, Journal of the Optical Society of America B, 39(3), pp. UPT1-UPT2.

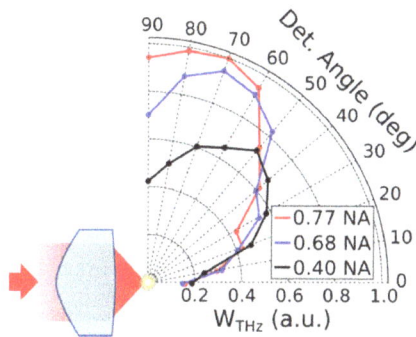

▲ Fig. 1.4.   THz pulse energy as a function of detection angle with three different achromatic lenses: 0.77 NA (red); 0.68 NA (blue); 0.40 NA (black). Reprinted from Ref. [18] with permission.

required to generate THz waves from such a microplasma is many orders of magnitude smaller compared to that of a filament. Also, there are substantial differences between "elongated" plasma and microplasma.

Fig. 1.4 shows the THz radiation pattern from a microplasma produced by three different high NA objectives, in which W_{THz} is the integrated THz pulse energy of a THz waveform measured from electro-optic sampling at different angles for detection [18]. The results show that higher NA gives

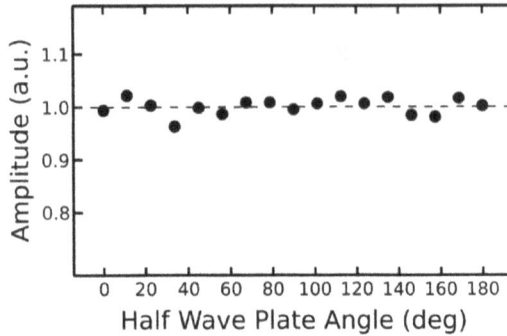

▲ Fig. 1.5. The dependence of THz field amplitude on the half-wave plate angle in the pump beam. Reprinted from Ref. [18] with permission.

a shorter plasma, leading to a larger angle from the optical axis for the maximum THz signal. With such a small longitudinal length of plasma, the THz emission direction is nearly perpendicular with respect to the optical axis of laser propagation; therefore, it is spatially separated from the residual laser excitation. The sideway THz radiation is attributed to a steep ponderomotive potential at the focal plane, which accelerates the free electrons created by photoionization.

For THz emission under ponderomotive force, the amplitude of THz wave only depends on the gradient of laser intensity. The polarization of the laser does not play a role in the emission process, which has been confirmed experimentally both in the elongated and microplasmas. Fig. 1.5 shows the dependence of THz field on the laser polarization, in which a half-wave plate is used to rotate the polarization of the pump beam. Moreover, the polarization of generated THz field remains the same for different polarization excitation.

1.3 Four-Wave Mixing and Transient Current Model

Under two-color excitation as shown in Fig. 1.2(b), THz generation efficiency is improved by three orders in the electric field, which attracts many researchers' attention for theoretical research. The third-order nonlinear effect of four-wave mixing is used to explain the high THz field emission from air plasma as the absence of second-order nonlinearity in gas media with a centrosymmetric symmetry [7], in which THz field amplitude is given by

$$E_{THz} \propto \chi^{(3)} E_\omega E_\omega E_{2\omega} \cos\varphi \qquad (1.2)$$

where E_ω and $E_{2\omega}$ are optical field amplitudes of the fundamental and second harmonic beams, respectively, φ is the relative phase of two beams, and $\chi^{(3)}$ is the third-order susceptibility. The observation of a high THz field indicates a large $\chi^{(3)}$ of gas plasma, which has been under debate for a long time due to the mystery of its origin. Kim *et al.* [24] attribute such a large susceptibility to a transient current produced by electron acceleration under an unsymmetric, time-dependent two-color laser field. The direction of the net polarization is determined by the relative phase of two beams. This current inside plasma is the microscopic source of the observed nonlinearity. Fig. 1.6 shows a typical THz waveform generated from a two-color plasma. Compared to single-color plasma, the THz signal from two-color shows a strong field with a broad bandwidth covering the whole THz range.

It is worth noting that both four-wave mixing and transient current models are intuitive and easy to implement and understand. However, for

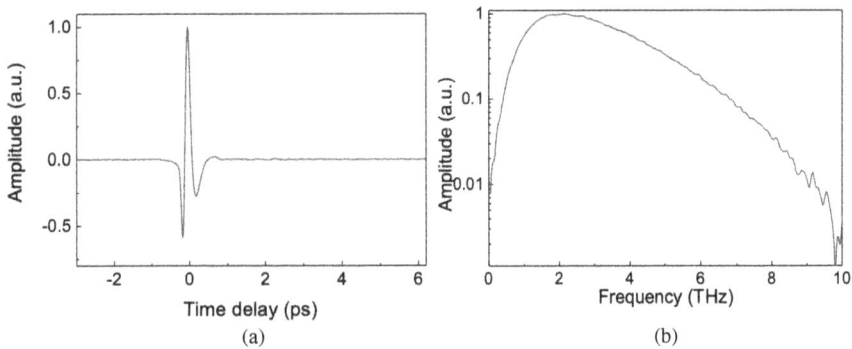

▲ Fig. 1.6. (a) A typical THz waveform generated and detected in dry nitrogen by a two-color plasma; (b) its spectrum. Reprinted with permission from Lu X. & Zhang X.-C. (2014). Investigation of ultra-broadband terahertz time-domain spectroscopy with terahertz wave gas photonics, Frontiers in Optoelectronics, 7, pp. 121–155.

a comprehensive description of THz emission from two-color air plasma, a numerical solution of the time-dependent Schrödinger equation and a time-space resolved three-dimensional simulation are needed.

1.4 THz Radiation from Relativistic Laser-Produced Plasmas

When a high-intensity laser pulse ($I > 10^{18}$ W/cm^2) is focused onto a target, relativistic electrons are produced with the velocity approaching the speed of light [25, 26]. Being accelerated by the laser wakefield, self-trapped electron bunches are generated with a longitudinal dimension in the order of the laser pulse duration. Such a short electron bunch can generate intense THz radiation through a variety of mechanisms, such as synchrotron radiation [27], sheath field acceleration [28], diffraction radiation (Smith-Purcell effect) [29], and transition radiation [30]. Generally, these THz sources can generate intense THz radiation with pulse energy several orders greater than that from traditional sources.

Electromagnetic radiation is emitted when charged particles pass through a boundary between two different media, which is known as transition radiation, as shown in Fig. 1.7. This contrasts with Cherenkov radiation, which occurs when charged particles pass through a homogeneous dielectric medium at a speed greater than the phase velocity

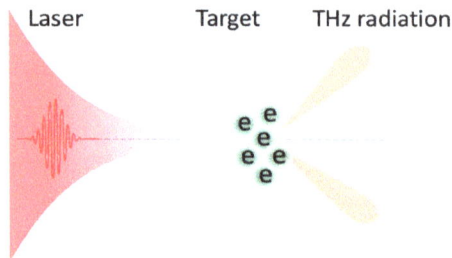

Laser   Target   THz radiation

▲ Fig. 1.7. Schematic of THz transition radiation production by a relativistic electron bunch crossing the plasma-vacuum interface.

of electromagnetic waves in that medium. Transition radiation is the main mechanism of THz emission if no magnetic field is applied. When the bunch length is smaller than the wavelength of interest, all the waves are generated in phase and coherent transition radiation is obtained. Unlike incoherent radiation, which scales linearly with the electron density, coherent transition radiation shows a quadratic dependence [31]. Typically, coherent transition radiation can be several orders higher than incoherent radiation since the number of electrons is about $10^{7\sim10}$ per bunch. The wavelength of the coherent transition radiation is limited by the transverse dimension (radiation diffraction) and the length of the electron bunch, which can be simplified in the following equation [32]:

$$2\pi\sigma_z < \lambda < \frac{2\pi R}{\sqrt{\gamma^2 - 1}} \tag{1.3}$$

where σ_z and R are the longitudinal length of the electron bunch and the target transverse dimension, respectively, and γ is the Lorentz factor of the relativistic electron. The relativistic electrons are very sensitive to the laser parameters, such as intensity and pulse duration. Therefore, the emitted THz radiation can be shaped by using a special target design and tuning the laser properties. After electrons escape from the rear side of the target, an electric field (sheath field) is created with the direction normal to the target surface. The ions are accelerated in this sheath field, which also could generate THz radiation.

Relativistic THz-matter interactions open a new research field. These extremely intense THz sources are useful not only for the generation or control of charged particle beams. The precise characterization of THz radiation from relativistic electrons can also be an efficient tool for diagnosing the process of laser-plasma interaction.

References

1. Brodeur A. & Chin S. L. (1998). Band-gap dependence of the ultrafast white-light continuum, Physical Review Letters, 80, pp. 4406–4409.

2. Brodeur A. & Chin S. L. (1999). Ultrafast white-light continuum generation and self-focusing in transparent condensed media, Journal of the Optical Society of America B, 16, pp. 637–650.
3. Couairon A. & Mysyrowicz A. (2007). Femtosecond filamentation in transparent media, Physics Reports, 441, pp. 47–189.
4. Chin S. L. (2010). *Femtosecond Laser Filamentation*. Springer.
5. Keldysh L. V. (1965). Ionization in the field of a strong electromagnetic wave, Soviet Physics — Journal of Experimental and Theoretical Physics, 20, pp. 1307–1314.
6. Hamster H., Sullivan A., Gordon S., White W. & Falcone R. W. (1993). Subpicosecond, electromagnetic pulses from intense laser-plasma interaction, Physical Review Letters, 71, pp. 2725–2728.
7. Cook D. J. & Hochstrasser R. M. (2000). Intense terahertz pulses by four-wave rectification in air, Optics Letters, 25, pp. 1210–1212.
8. Zhang X.-C., Shkurinov A. & Zhang Y. (2017). Extreme terahertz science, Nature Photonics, 11, pp. 16–18.
9. Xie X., Dai J. & Zhang X.-C. (2006). Coherent control of THz wave generation in ambient air, Physical Review Letters, 96, pp. 075005.
10. Dai J., Karpowicz N. & Zhang X.-C. (2009). Coherent polarization control of terahertz waves generated from two-color laser-induced gas plasma, Physical Review Letters, 103, pp. 023001.
11. Wen H. & Lindenberg A. M. (2009). Coherent terahertz polarization control through manipulation of electron trajectories, Physical Review Letters, 103, pp. 023902.
12. Clough B., Liu J. & Zhang X.-C. (2011). "All air-plasma" terahertz spectroscopy, Optics Letters, 36, pp. 2399–2401.
13. Dai J., Clough B., Ho I. C., Lu X., Liu J. & Zhang X.-C. (2011). Recent progresses in terahertz wave air photonics, IEEE Transactions on Terahertz Science and Technology, 1, pp. 274–281.
14. Dai J., Liu J. & Zhang X.-C. (2011). Terahertz wave air photonics: terahertz wave generation and detection with laser-induced gas plasma, IEEE Journal of Selected Topics in Quantum Electronics, 17, pp. 183–190.
15. Sagisaka A., Daido H., Nashima S., Orimo S., Ogura K., Mori M., Yogo A., Ma J., Daito I., Pirozhkov A. S., Bulanov S. V., Esirkepov T. Z., Shimizu K. & Hosoda M. (2008). Simultaneous generation of a

proton beam and terahertz radiation in high-intensity laser and thin-foil interaction, Applied Physics B, 90, pp. 373–377.

16. Liao G., Li Y., Liu H., Scott G. G., Neely D., Zhang Y., Zhu B., Zhang Z., Armstrong C. & Zemaityte E. (2019). Multimillijoule coherent tera-hertz bursts from picosecond laser-irradiated metal foils, Proceedings of the National Academy of Sciences, 116, pp. 3994–3999.

17. Boot H. A. H. & Harvie R. B. R.-S. (1957). Charged particles in a non-uniform radio-frequency field, Nature, 180, pp. 1187–1187.

18. Buccheri F. & Zhang X.-C. (2018). Generation and detection of pulsed terahertz waves in gas: from elongated plasmas to microplasmas, Frontiers of Optoelectronics, 11, pp. 209–244.

19. D'Amico C., Houard A., Franco M., Prade B., Mysyrowicz A., Couairon A. & Tikhonchuk V. T. (2007). Conical forward THz emission from femtosecond-laser-beam filamentation in air, Physical Review Letters, 98, pp. 235002.

20. Mlejnek M., Wright E. M. & Moloney J. V. (1998). Dynamic spa-tial replenishment of femtosecond pulses propagating in air, Optics Letters, 23, pp. 382–384.

21. Talebpour A., Petit S. & Chin S. L. (1999). Re-focusing during the propagation of a focused femtosecond Ti:Sapphire laser pulse in air, Optics Communications, 171, pp. 285–290.

22. Buccheri F. & Zhang X.-C. (2015). Terahertz emission from laser-induced microplasma in ambient air, Optica, 2, pp. 366–369.

23. Buccheri F., Liu K. & Zhang X.-C. (2017). Terahertz radiation enhanced emission of fluorescence from elongated plasmas and microplasmas in the counter-propagating geometry, Applied Physics Letters, 111, pp. 091103.

24. Kim K.-Y., Glownia J. H., Taylor A. J. & Rodriguez G. (2007). Terahertz emission from ultrafast ionizing air in symmetry-broken laser fields, Optics Express, 15, pp. 4577–4584.

25. Malka V., Fritzler S., Lefebvre E., Aleonard M.-M., Burgy F., Chambaret J.-P., Chemin J.-F., Krushelnick K., Malka G. & Mangles S. (2002). Electron acceleration by a wake field forced by an intense ultrashort laser pulse, Science, 298, pp. 1596–1600.

26. Mangles S. P., Murphy C., Najmudin Z., Thomas A. G. R., Collier J., Dangor A. E., Divall E., Foster P., Gallacher J. & Hooker C. (2004).

Monoenergetic beams of relativistic electrons from intense laser–plasma interactions, Nature, 431, pp. 535–538.

27. Evain C., Szwaj C., Roussel E., Rodriguez J., Le Parquier M., Tordeux M.-A., Ribeiro F., Labat M., Hubert N. & Brubach J.-B. (2019). Stable coherent terahertz synchrotron radiation from controlled relativistic electron bunches, Nature Physics, 15, pp. 635–639.

28. Gopal A., Herzer S., Schmidt A., Singh P., Reinhard A., Ziegler W., Brömmel D., Karmakar A., Gibbon P., Dillner U., May T., Meyer H. G. & Paulus G. G. (2013). Observation of gigawatt-class THz pulses from a compact laser-driven particle accelerator, Physical Review Letters, 111, pp. 074802.

29. Yi L. & Fülöp T. (2019). Coherent diffraction radiation of relativistic terahertz pulses from a laser-driven microplasma waveguide, Physical Review Letters, 123, pp. 094801.

30. Liao G.-Q., Li Y.-T., Zhang Y.-H., Liu H., Ge X.-L., Yang S., Wei W.-Q., Yuan X.-H., Deng Y.-Q., Zhu B.-J., Zhang Z., Wang W.-M., Sheng Z.-M., Chen L.-M., Lu X., Ma J.-L., Wang X. & Zhang J. (2016). Demonstration of coherent terahertz transition radiation from relativistic laser-solid interactions, Physical Review Letters, 116, pp. 205003.

31. Leemans W., Geddes C., Faure J., Tóth C., Van Tilborg J., Schroeder C., Esarey E., Fubiani G., Auerbach D. & Marcelis B. (2003). Observation of terahertz emission from a laser-plasma accelerated electron bunch crossing a plasma-vacuum boundary, Physical Review Letters, 91, pp. 074802.

32. Schroeder C. B., Esarey E., van Tilborg J. & Leemans W. P. (2004). Theory of coherent transition radiation generated at a plasma-vacuum interface, Physical Review E, 69, pp. 016501.

© 2024 World Scientific Publishing Company
https://doi.org/10.1142/9789811265648_0002

Chapter 2

The Basics of Broadband Terahertz Wave Generation

from Liquids

Yiwen E, X.-C. Zhang

The Institute of Optics, University of Rochester, USA

Terahertz (THz) liquid sources were conspicuously absent until very recently. It is well known that water, the most common liquid, is a strong absorber in the THz frequency range [1–3]. Therefore, liquid water has historically been sworn off as a source for THz emission. In 2017, broadband THz wave generation from a gravity-driven, free-flowing water film [4] and from liquids in a cuvette [5] has been experimentally demonstrated. Since then, different liquids have been investigated for THz emission. Fig. 2.1 shows different liquid targets, including water [6–10], liquid nitrogen [11, 12], and liquid gallium [13–15]. In this chapter, we will discuss the basics of THz wave generation from liquids.

Most liquids with polar molecules strongly absorb far-infrared electromagnetic wave, including THz waves. Water vapor has many sharp absorption lines in the THz range resulting from molecular vibrational and rotational modes, and liquid water has a continuous absorption covering the THz regime, with its power absorption coefficient $\alpha = 220$ cm^{-1} at 1 THz [3]. If N_{out} and N_{in} are numbers of input and output THz photons,

Water film	Water line	Liquid nitrogen	Liquid gallium

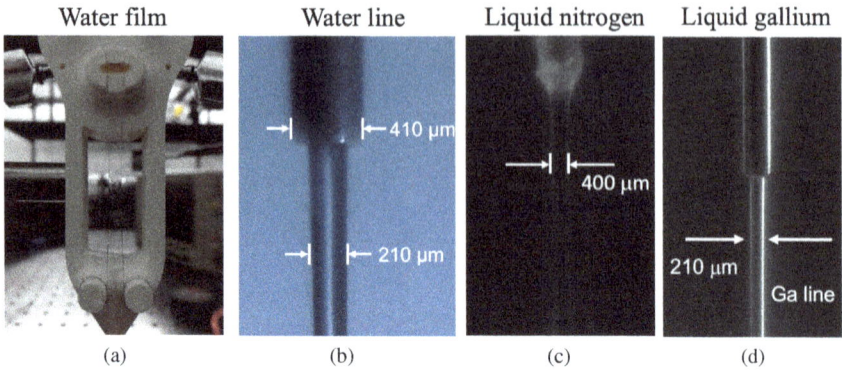

(a)	(b)	(c)	(d)

▲ Fig. 2.1.   Different liquid targets used for THz emission: (a) water film, (b) water line, (c) liquid nitrogen, and (d) liquid metal.

▼ Table 2.1.   Input photon numbers to get one photon out from a water film with thickness d.

Water thickness, d (mm)	THz photons in, N_{in} (#)	THz photons out, N_{out} (#)
1	3.6×10^9	1
0.1	9	1

respectively, they follow Eq. (2.1). Here, the reflection effect of interfaces is neglected.

$$N_{out} = e^{-\alpha d} N_{in} \tag{2.1}$$

Calculated through this equation, Table 2.1 lists numbers of THz photons needed in order to get one THz photon out after passing through a water film with a thickness of d mm. For a 1 mm-thick water layer, it requires 3.6 billion THz photons to get one THz photon out. This calculation does not include the air/water interface reflection. To reduce the attenuation of THz waves, a thinner water layer is the key. As an example, the number of N_{in} drops eight orders, from 3.6 billion input THz photons to 9 input THz photons, while the thickness d decreases by one order, from 1 mm to 0.1 mm.

There are various approaches to producing a thin water film. As shown in Fig. 2.1(a), a flowing water film with a thickness of 120 μm is guided by two parallel thin wires. Benefiting from the surface tension of water, the film thickness can be tuned by the separation of the wire gap as well as by increasing/decreasing the flow rate. The film transparency indicates surface smoothness. Also, a liquid jet shaped by a nozzle can be used to make a thin water film. The thickness of the flowing film can be measured through an optical autocorrelation system by inserting the film into one of the two laser beams. The signal peak in the time domain delays to a certain position compared to that of the air signal. The peak position moves linearly with the increase of the film thickness.

Fig. 2.2(a) shows a typical setup to study THz wave generation from liquids. A Ti:sapphire regenerative amplifier with mJ-level pulse energy is used for excitation. The laser beam is focused into a thin water film. Fig. 2.3 shows a photo of a water film ionized by 800 nm pulses, in which a parabolic mirror with one-inch focal length is used for excitation. Compared to a gas material, liquid has a relatively lower ionization threshold [16–18], leading to a lower intensity in the requirement for creating a plasma in a liquid material. After the target, the THz wave is collected by a pair of parabolic mirrors and detected through either electro-optic sampling or a THz energy

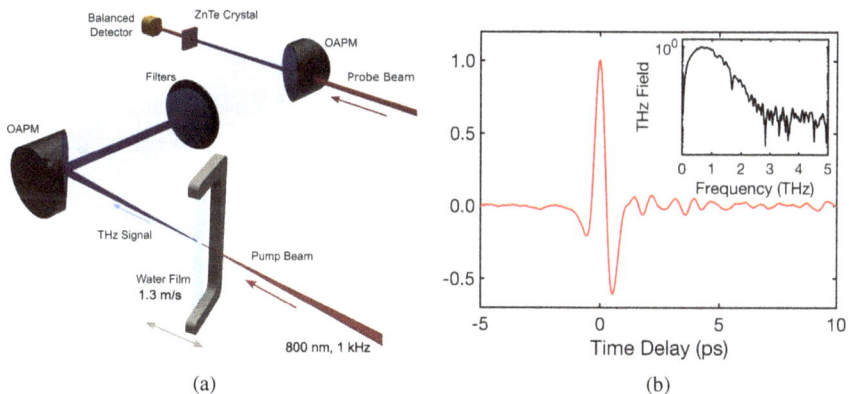

(a)   (b)

▲ Fig. 2.2.   (a) A typical system to study THz emission from a water target. The laser pulses with a central wavelength of 800 nm and a repetition rate of 1 kHz are focused on the liquid target. The THz wave is detected through electro-optic sampling. Reprinted from Ref. [4] with permission. (b) THz waveform generated from a liquid water target. The inset shows its spectrum. Reprinted from Ref. [6] with permission.

▲ Fig. 2.3. The bright spot is a plasma in the water film produced by 800 nm laser pulses. The white fluorescence spot from a water plasma is much stronger than that from an air plasma due to the much higher molecular density of liquid material contributing to the light.

detector. For a flowing liquid target, a certain flow rate is needed to assure that each pulse interacts with a fresh area with no turbulence caused by the previous pulse. Fig. 2.2(b) shows the THz waveform generated from a thin flowing water target, which is detected by a 2 mm ZnTe crystal. The inset shows its corresponding spectrum. The central frequency of the THz pulse is around 0.5 THz. The optimized signal is measured when the angle of incidence is 65° for the film thickness of 180 μm [6]. Compared with the THz pulse generated from gas, the water signal shows more low-frequency components and less high-frequency components, which is attributed to the strong absorption of high frequency in water. For a liquid with low absorption, a broader bandwidth is expected.

The white spot from a liquid plasma as shown in Fig. 2.3 is fluorescence, which is caused by the radiative relaxation of excited electrons. In ambient air, the main contributions to fluorescence are nitrogen molecular lines, atomic lines, and a broad continuum extending from UV to infrared due to electron recombination. In the case of loose focusing, the molecular and atomic lines are relatively strong. In tight focusing with higher laser intensity, the continuum emission increases greatly, and the lines disappear [19]. For the liquid case, the continuum emission is always high since the molecular fluorescence emission is short-lived because of the strong collision in liquid with a much higher molecular density [20]. A bright white spot is always observed in a liquid case with laser excitation.

Fig. 2.4 plots THz waveforms generated from water films with different thicknesses. The peak position of each waveform is projected in the horizontal plane, which linearly shifts in time with increased thickness. This time shift comes from the path difference of THz waves in water with different thicknesses. By fitting the shifts in time, shown as a red dashed line in the plot, the refractive index of water is obtained as n_{THz} = 2.29, if we assume that increasing the thickness of the water film goes symmetrically with respect to the center of the film. This number agrees with the result measured in previous works, in which n_{THz} is 2.27 at 0.5 THz. The attenuation of the THz emission with the increased

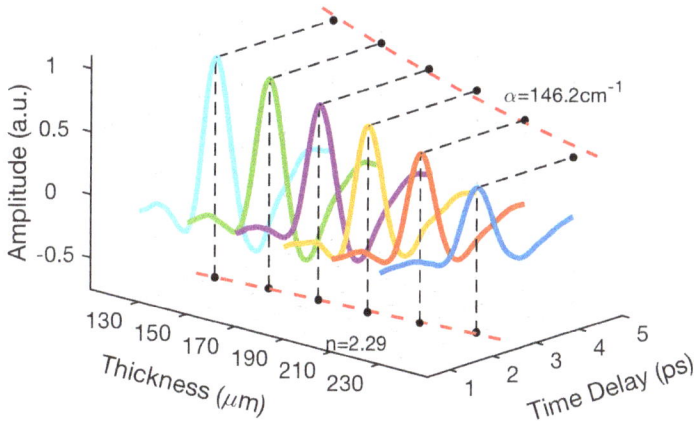

▲ Fig. 2.4.   THz wave generation from water films with different thicknesses. The refractive index of water at 0.5 THz is calculated to be 2.29 from the time shift of the THz field. The absorption coefficient of water at 0.5 THz is calculated to be 146.2 cm⁻¹ from the attenuation of the THz field's amplitude. Reprinted from Ref. [24] with permission.

thickness is also shown in this plot, which results from the increased absorption of a thicker water film. The peak field of each waveform is projected in the vertical plane. From an exponential fitting, shown as a red dashed line in the plot, the absorption coefficient of water is calculated as $\alpha = 146.2$ cm^{-1}. It should be noted that the refractive index and absorption are obtained by fitting the tendency of the peak field of the THz waveform, which is an averaged result of a broad bandwidth. A high yield of THz emission is expected if an ultrathin target is used.

Comparing liquid with gas targets, several differences are involved in the interaction process. The liquid target introduces interfaces in both ionization and emission processes, which brings a new parameter to be considered and controlled. For a film target, we found that there is an optimal angle of incidence for the laser excitation. As shown in Fig. 2.5(a), when the film at the focus is rotated, the incident angle of the laser is adjusted accordingly. Fig. 2.5(b) shows the dependence of THz pulse energy on the incident angle. There is little signal for the case of normal incidence. The energy goes up by increasing the angle of incidence both in clockwise and counterclockwise directions. The optimized angle is related to the thickness of the film. For a 120 μm-thick water film, the optimized

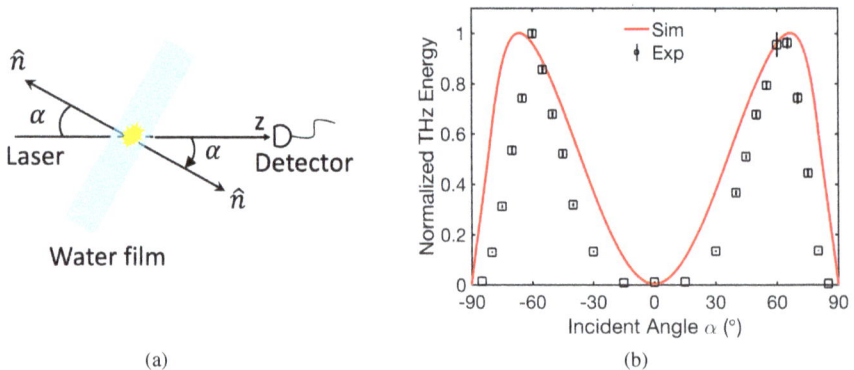

(a)   (b)

▲ Fig. 2.5.   (a) Illustration of the incident angle of laser for excitation. The detector is in the direction of laser propagation. The incident angle is adjusted by rotating the water film. $\hat{n}$ is the surface normal of the water film. (b) The dependence of THz pulse energy on the angle of incidence of the laser beam. Reprinted from Ref. [6] with permission.

angle for the detection in the laser propagation direction is ±65°, where the sign ± indicates the angles of opposite directions.

To understand the optimized angle of incidence, several factors need to be considered, including laser-induced transient current for radiating THz wave, THz propagation in the liquid, and the influence of the detection direction. When the laser beam is incident on the film target, the beam refracts at the air/liquid interface. According to Snell's law and the Fresnel equations, the refractive angle θ_r and the transmittance T_1 for the laser beam with an angle of incidence α at the first air/water interface are:

$$\theta_r(\alpha) = \arcsin\left(\frac{n_A}{n_L}\sin\alpha\right) \tag{2.2}$$

$$T_1(\alpha) = 1 - \left|\frac{n_A\cos\theta_r - n_L\cos\alpha}{n_A\cos\theta_r + n_L\cos\alpha}\right|^2 \tag{2.3}$$

where n_A and n_L are the refractive indices of the 800 nm optical beam in air and liquid, respectively. According to Eq. (2.2), the refractive angle θ_r is 48.8° for the case of water when α is 90°.

Since the ionization potential threshold of air is usually lower than that of liquid, ionization may happen at the interface if the laser intensity is greater than the threshold of liquid but lower than that of air. Wherever the front of the optical pulse reaches the ionization threshold, electrons and ions are produced through multiphoton ionization, tunneling ionization, and cascade ionization. Initially, these charges are distributed homogeneously, which means there is no difference between the electron and ion distributions.

Under the ponderomotive force, a net charge density is produced, where electrons are accelerated toward the area of lower laser intensity. The separated distributions between electrons and ions act like a Hertzian dipole. When the laser propagates away from plasma, the restoring force pulls the electrons back to ions, which is similar to the case of THz emission from single-color air plasma. Therefore, a transient current is created

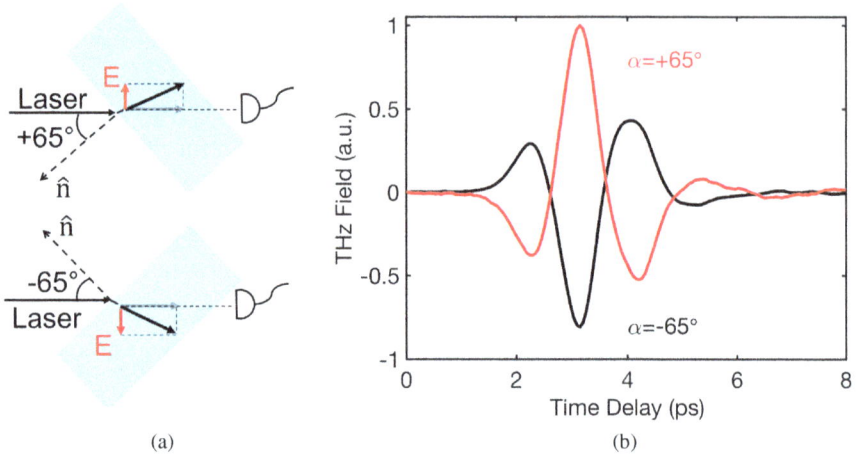

▲ Fig. 2.6.   Flipped THz waveforms under opposite angle incidence are plotted in (b). The projection of the dipole in the direction perpendicular to the laser propagation is measured by the THz field detector, which shows opposite directions for the projected dipole perpendicular to the laser propagation direction (red arrow) for two cases plotted in (a). Reprinted from Ref. [23] with permission.

under the ponderomotive and restoring force with a lifetime of the optical pulse duration, which radiates broadband THz pulse. Two observations are supporting the discussion of the dipole direction, which is along the laser propagation direction. One is the little THz energy for the case of laser normal incidence. The other one is the flipped waveforms measured at the opposite angles of incidence, as shown in Fig. 2.6.

Fig. 2.6 shows the flipped THz waveforms. Assuming that the incident laser propagates in the z direction, a transient dipole is created along the refracted laser beam after the air/liquid interface. According to the radiation pattern of a dipole, the detector in the z direction detects the energy from the projected dipole, which is perpendicular to the z direction, shown as a red arrow in Fig. 2.6(a). For the case of opposite angles of incidence, the projected dipole shows an opposite polarity, leading to a flipped THz waveform in the far field. Fig. 2.6(b) shows THz waveforms measured at ±65°, which have the same field strength with opposite polarities. Moreover, THz emission refracts at the liquid/air surface and propagates in the far field. For a flat surface, some THz waves with a large angle of incidence on the rear surface of the target experience total internal reflection. Therefore, the total output THz energy from a liquid target is decided by the optical refraction at the air/liquid interface, radiation pattern from the transient dipole, and THz refraction at the liquid/air interface. In the case of a water film, about 80% of the total energy is dissipated even at the optimal angle of incidence due to the strong absorption and total internal reflection. To improve the THz energy in the far field, a target geometry could be designed to prevent the effect of total internal reflection. By using a liquid with low absorption at THz range, it is expected that the THz pulse energy will be improved by a lot.

Although the THz emission mechanism from liquid under single-color excitation is similar to that of air plasma, there is a big difference in the ionization process of liquid and gas media, which is revealed by the study of THz emission from liquids. One of the most important observations of THz wave emission from a liquid is the preference of long pulse duration for excitation, which is very counterintuitive. Fig. 2.7 shows the dependence

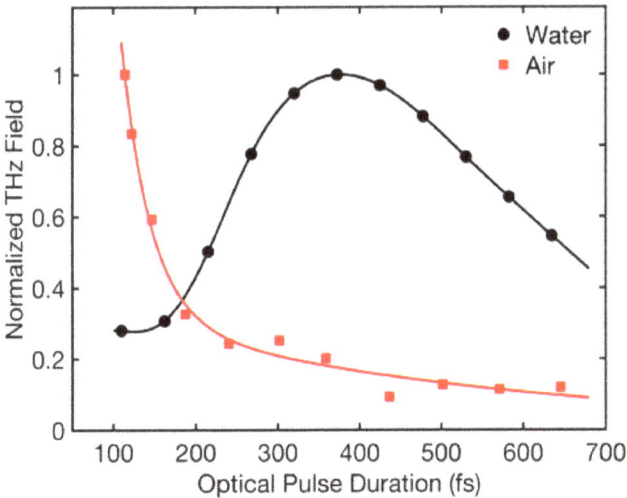

▲ Fig. 2.7.   THz field strength *vs* the optical pulse duration in water (black) and air (red), respectively.

of THz field on the optical pulse duration for air and liquid water. For air plasma, the strongest signal is always obtained when the optical pulse is the shortest since it provides the highest peak intensity for ionization. However, for all the liquids that have been tested in our lab, the THz field increases by increasing optical pulse duration from the minimum pulse duration (say 35 fs from a typical laser) until it reaches a peak at a longer pulse duration. For a regenerative amplifier laser, the pulse duration is tuned by moving the grating in the compressor. Since this method changes the chirp of the pulse, both positive and negative chirps are tested and they generate a stronger signal with a similar spectrum, which shows that the pulse duration of the laser pulse is more important rather than the chirp.

This observation shows that the ionization process of liquid is different from that of gas. For a general ionization process, electrons are created first through multiphoton ionization or tunnel ionization. These electrons as seeds will be accelerated under an intense laser field. More electrons will be ionized through cascade ionization by colliding with high-energy electrons. For air plasma, the mean free time of electrons is about 100 fs (Table 2.2), which is longer or comparable to the optical pulse duration for excitation. In this case, collision barely happens. However, the mean free time of electrons in liquid is about or lower than 1 fs, which is much smaller than the optical pulse duration. Therefore, cascade ionization plays a more important role in the liquid ionization process. The calculation shows a similar tendency of electron density to the THz pulse energy, which explains the preference of a long pulse for excitation in liquids.

To boost the THz emission efficiency from a liquid target, an asymmetric excitation scheme with two-color pulses is employed, which

▼ Table 2.2. A comparison of ionization potential and mean free time of electron for water and air.

	Water	Air
Ionization potential	~6.5 eV	12 eV
Mean free time	~1 fs	~100 fs

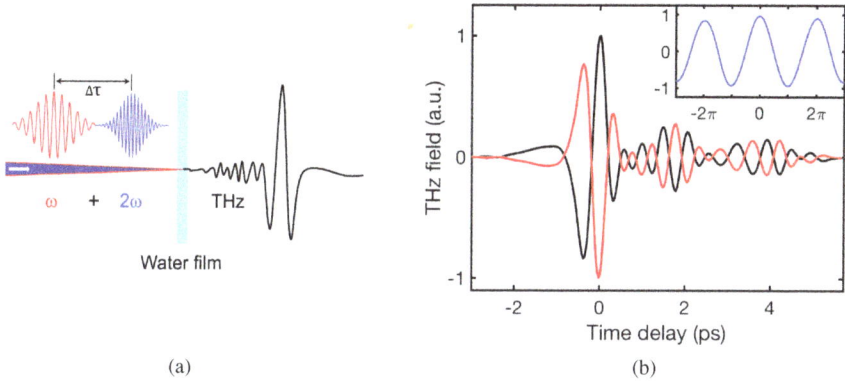

▲ Fig. 2.8. (a) The schematic diagram of THz emission under two-color excitation. (b) Two THz waveforms show opposite polarity by changing the relative phase between two color pulses with a π shift. The inset shows the dependence of THz peak field on the relative phase. Reprinted from Ref. [22] with permission.

has previously shown a three-order enhancement in the electric field for gas plasma. Both the fundamental and second harmonic beams are focused to create the plasma in liquid, in which the optical fields are asymmetric since the relative phase is between two beams. The schematic diagram of two-color excitation in water film is shown in Fig. 2.8(a). An in-line phase compensator is usually used to control the relative phase between two beams. Compared with single-color excitation, a 10-time increase in field (100 times in energy) is obtained with the two-color excitation. A possible reason for the relatively low enhancement compared with the gas case is the quasi-free electrons ionized in liquid, since most electrons are excited to the continuum band rather than free space. Fig. 2.8(b) shows that the polarity of the THz electric field is completely flipped over by changing the relative phase between the fundamental and the second harmonic pulses by π. The inset of Fig. 2.8(c) plots the THz field as a function of optical phase delay, which indicates that the polarity of the THz electric field is gradually changed with the phase delay.

It is well known that liquid water has the highest density at 3.98°C [21]. Fig. 2.9(a) shows the dependence of water density on temperature in the range of 0–10°C. The density difference over this range is about

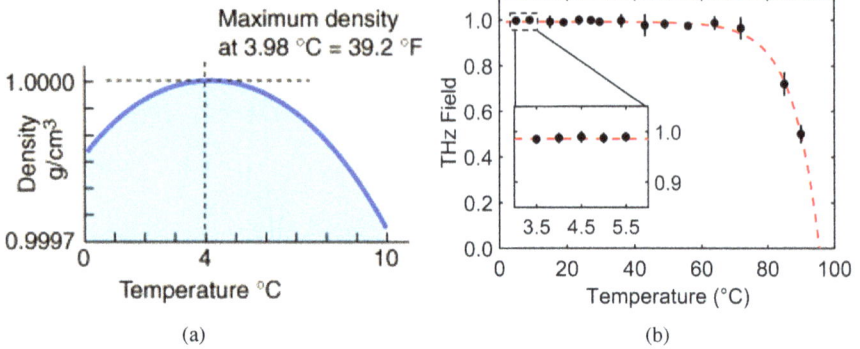

▲ Fig. 2.9. (a) The dependence of water density on temperature (from Wikipedia). (b) The dependence of THz peak field on the water temperature. Reprinted with permission from Cao Y., E Y., Tcypkin A., Huang P. & Zhang X.-C. (2020). Terahertz wave generation from water at different temperatures, 45th International Conference on Infrared, Millimeter, and Terahertz Waves (IRMMW-THz), Buffalo, NY, USA, pp. 1–2.

0.003%. Preliminary measurement of THz signal from water at different temperatures is plotted in Fig. 2.9(b). In order to see the effect at 4°C, assuming the THz emission is linearly related to the target density, then the control step of water temperature should be less than 0.1°C, and the signal-to-noise ratio of THz field measurement should be better than 10,000. This requires a further improvement on our liquid-target THz system to measure a small change in the THz field. A fast drop of the THz signal when the temperature is greater than 75°C might be due to the water vapor generated by the laser beam.

In summary, this chapter focuses on the basics of using liquids as targets to generate THz radiation with laser excitation. The discussion provides the idea of developing liquid-based intense THz sources and also contributes to enthralling insights into the study of laser-liquid interaction. Broadband THz wave generation from a water film was first demonstrated experimentally. The mechanism of the generation process is attributed to laser-induced plasma formation in water associated with a ponderomotive force-induced dipole radiation model. Specifically, it is observed that THz radiation from liquid prefers long optical pulses for excitation. Under the two-color excitation scheme, the enhancement of THz emission from a water film was also confirmed [22]. It needs to be highlighted that THz wave generation from water is stronger than that from air with the one-color excitation scheme, which is coincident with expectation from the higher molecular density of water. However, water is no longer the winner with the two-color excitation scheme. In water, cascade ionization favors a long pulse duration to exponentially produce electrons. But the second harmonic generation efficiency from the fundamental laser pulse significantly drops as the optical pulse duration increases. The discrepant demand regarding the pulse duration significantly limits the generation of THz waves under the two-color excitation. For more details on THz emission from liquids, the reader is advised to read the references [23–25]. From the perspective of the development of intense liquid THz sources, the two-color excitation scheme may still be the way to go. Wise management of laser pulse duration is imperative. It should be noticed that only a few types of liquids have been tested so far. The best liquid source for THz radiation may still be out there awaiting discovery, so systematic tests of other liquids are necessary. Commercializing a compact liquid THz source is the endgame.

References

1. Thrane L., Jacobsen R. H., Uhd Jepsen P. & Keiding S. R. (1995). THz reflection spectroscopy of liquid water, Chemical Physics Letters, 240, pp. 330–333.
2. Rønne C., Thrane L., Åstrand P.-O., Wallqvist A., Mikkelsen K. V. & Keiding S. r. R. (1997). Investigation of the temperature dependence of dielectric relaxation in liquid water by THz reflection spectroscopy and molecular dynamics simulation, The Journal of Chemical Physics, 107, pp. 5319–5331.
3. Wang T., Klarskov P. & Jepsen P. U. (2014). Ultrabroadband THz time-domain spectroscopy of a free-flowing water film, IEEE Transactions on Terahertz Science and Technology, 4, pp. 425–431.
4. Jin Q., E Y., Williams K., Dai J. & Zhang X.-C. (2017). Observation of broadband terahertz wave generation from liquid water, Applied Physics Letters, 111, pp. 071103.

5. Dey I., Jana K., Fedorov V. Y., Koulouklidis A. D., Mondal A., Shaikh M., Sarkar D., Lad A. D., Tzortzakis S., Couairon A. & Kumar G. R. (2017). Highly efficient broadband terahertz generation from ultra-short laser filamentation in liquids, Nature Communications, 8, pp. 1184.

6. E Y., Jin Q., Tcypkin A. & Zhang X.-C. (2018). Terahertz wave generation from liquid water films via laser-induced breakdown, Applied Physics Letters, 113, pp. 181103.

7. E Y., Jin Q. & Zhang X.-C. (2019). Enhancement of terahertz emission by a preformed plasma in liquid water, Applied Physics Letters, 115, pp. 101101.

8. Tcypkin A. N., Ponomareva E. A., Putilin S. E., Smirnov S. V., Shtumpf S. A., Melnik M. V., E Y., Kozlov S. A. & Zhang X.-C. (2019). Flat liquid jet as a highly efficient source of terahertz radiation, Optics Express, 27, pp. 15485–15494.

9. Zhang L.-L., Wang W.-M., Wu T., Feng S.-J., Kang K., Zhang C.-L., Zhang Y., Li Y.-T., Sheng Z.-M. & Zhang X.-C. (2019). Strong terahertz radiation from a liquid-water line, Physical Review Applied, 12, pp. 014005.

10. Jin Q., E Y., Gao S. & Zhang X.-C. (2020). Preference of subpicosecond laser pulses for terahertz wave generation from liquids, Advanced Photonics, 2, pp. 015001.

11. Balakin A. V., Coutaz J.-L., Makarov V. A., Kotelnikov I. A., Peng Y., Solyankin P. M., Zhu Y. & Shkurinov A. P. (2019). Terahertz wave generation from liquid nitrogen, Photonics Research, 7, pp. 678–686.

12. E Y., Cao Y., Ling F. & Zhang X.-C. (2020). Flowing cryogenic liquid target for terahertz wave generation, AIP Advances, 10, pp. 105119.

13. Cao Y., E Y., Huang P. & Zhang X.-C. (2020). Broadband terahertz wave emission from liquid metal, Applied Physics Letters, 117, pp. 041107.

14. Solyankin P. M., Lakatosh B. V., Krivokorytov M. S., Tsygvintsev I. P., Sinko A. S., Kotelnikov I. A., Makarov V. A., Coutaz J.-L., Medvedev V. V. & Shkurinov A. P. (2020). Single free-falling droplet of liquid metal as a source of directional terahertz radiation, Physical Review Applied, 14, pp. 034033.

15. Garriga Francis K., Cao Y., E Y., Ling F., Lim Pac Chong M. & Zhang X.-C. (2021). Forward terahertz wave generation from liquid

gallium in the non-relativistic regime, Journal of the Optical Society of America B, 38, pp. 3639–3645.

16. Williams F., Varma S. & Hillenius S. (1976). Liquid water as a lone-pair amorphous semiconductor, The Journal of Chemical Physics, 64, pp. 1549–1554.

17. Nikogosyan D. N., Oraevsky A. A. & Rupasov V. I. (1983). Two-photon ionization and dissociation of liquid water by powerful laser UV radiation, Chemical Physics, 77, pp. 131–143.

18. Crowell R. A. & Bartels D. M. (1996). Multiphoton ionization of liquid water with 3.0–5.0 eV photons, The Journal of Physical Chemistry, 100, pp. 17940–17949.

19. Buccheri F., Huang P. & Zhang X.-C. (2018). Generation and detection of pulsed terahertz waves in gas: from elongated plasmas to microplasmas, Frontiers of Optoelectronics, 11, pp. 209–244.

20. Chin S.L. (2010). *Femtosecond Laser Filamentation*. Springer.

21. Franks F. (1967). *Physico-chemical Processes in Mixed Aqueous Solvents*. Elsevier.

22. Jin Q., Dai J. M., E Y. & Zhang X.-C. (2018). Terahertz wave emission from a liquid water film under the excitation of asymmetric optical fields, Applied Physics Letters, 113, pp. 261101.

23. E Y., Zhang L., Tcypkin A., Kozlov S., Zhang C. & Zhang X.-C. (2021). Broadband THz sources from gases to liquids, Ultrafast Science, 2021, pp. 9892763.

24. Jin Q., E Y. & Zhang X.-C. (2021). Terahertz aqueous photonics, Frontiers of Optoelectronics, 14, pp. 37–63.

25. E Y., Zhang L., Tsypkin A., Kozlov S., Zhang C. & Zhang X.-C. (2022). Progress, challenges, and opportunities of terahertz emission from liquids, Journal of the Optical Society of America B, 39, pp. A43–A51.

© 2024 World Scientific Publishing Company
https://doi.org/10.1142/9789811265648_0003

Chapter 3

Plasma Generation in Liquid Water Under Intense Laser Irradiation

Jiyu Xu, Yunzhe Jia, Sheng Meng

Beijing National Laboratory for Condensed Matter Physics and Institute of Physics, Chinese Academy of Sciences, China

3.1 Introduction

Under intense laser irradiation, liquid water will be excited and ionized. The ionization of liquid water occurs at the femtosecond timescale, which leads to fast water dissociation and plasma generation. However, the underlying microscopic mechanisms of plasma generation in liquid water have been scarcely discussed despite many macroscopic experimental observations.

In this chapter, we introduce the microscopic mechanisms of laser-induced ultrafast plasma generation in liquid water. The separative evolutions of electronic and ionic subsystems immediately following intense laser irradiations are presented. In Section 3.2, an overview on the laser-induced macroscopic phenomena is briefly introduced. Then, the numerical simulations of electronic excitations are discussed in Section 3.3. In Section 3.4, the microscopic atomistic dynamics of liquid water following laser irradiation are shown. We complete the chapter with a presentation of the nonadiabatic *ab initio* simulations in Section 3.5 to give a complete description of laser-induced plasma generation in liquid water.

3.2 Macroscopic Observations of Laser-Induced Water Plasma

The dynamics of liquid water impacted by laser pulses range over eight orders of magnitude in time. The photoexcitation of liquid water takes place at the femtosecond timescale, and the consequent optical breakdown and water dissociation give rise to plasma generation in liquid water. The ensuing thermalization processes could give rise to intriguing macroscopic phenomena, such as shock-wave emission, spallation, and drastic explosion. The optical breakdown of liquid water is accompanied by plasma luminescence and followed by shock wave emission and cavitation for nanosecond and picosecond pulses, respectively, while optical breakdown is detected by the formation of a cavitation bubble for femtosecond pulses. Therefore, both plasma luminescence and formation of cavitation bubbles can be used as the experimental breakdown criteria of plasma generation in liquid water [1].

Schaffer et al. [2] observed that a hot electronic plasma with a radius of about 2 μm is formed within 200 fs following 100 fs, 800 nm laser irradiation (Fig. 3.1(a)). Then the plasma expands rapidly from 30 ps to 200 ps and the radius of the plasma begins to oscillate at an average of 6 μm. At about 800 ps a pressure wave separates from the central bubble and travels at the speed of sound in liquid water. At 10 ns the central bubble expands as a cavitation bubble, then reaches the maximum radius of 100 μm at 5 μs and re-collapses at 11 μs. On the other hand, laser-induced plasma generation and vaporization can lead to macroscopic motions in sessile drops and confinement geometries. Klein et al. [3] observed that a water drop hit by a mJ nanosecond laser pulse propels forward at the velocity of several meters per second and deforms into fragments (Fig. 3.1(b)), which are attributed to the recoil due to vaporization on the face of the water drop. Stan et al. [4] showed that the intense X-ray laser pulses induce violent explosions in water drops and jets (Fig. 3.1(c)). The explosive vaporization of a jet section near the X-ray spot leads to the formation of a gap, then liquid from the jet ends is pushed into thin conical films of water and later collapses onto the jet.

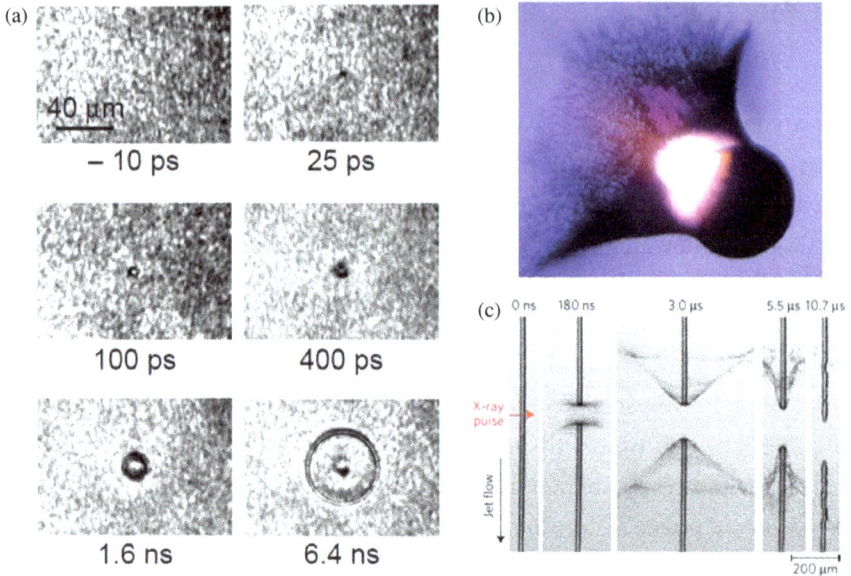

▲ Fig. 3.1.   The laser-induced macroscopic phenomena of water. (a) Images of femtosecond laser-induced breakdown of liquid water for various time delays. A 100 fs, 800 nm wavelength pump pulse is used. Reprinted from Ref. [2] with permission. (b) The 532 nm laser pulses impact from the left on a water drop of radius 0.9 mm, and the white plasma glow and violent ablation from the water drop are observed. Reprinted from Ref. [3] with permission. (c) Images of explosions induced by 0.75 mJ, 8.2 keV X-ray free electron laser pulses in a 20 μm diameter water jet. Reprinted from Ref. [4] with permission.

3.3 Numerical Simulations of Laser-Induced Plasma Generation

In numerical simulations [1], the breakdown threshold is defined as a critical free-electron density ρ_{cr} or the irradiance (energy) required to produce that free-electron density, and water dissociation is usually neglected to reduce complexities. The electron density threshold is expressed as

$$\rho_{cr} = \omega^2 m_c \varepsilon_0 / e^2 \tag{3.1},$$

where ω is the angular frequency of laser pulse, ε_0 denotes the vacuum dielectric permittivity, m_c denotes the effective mass of an electron in the

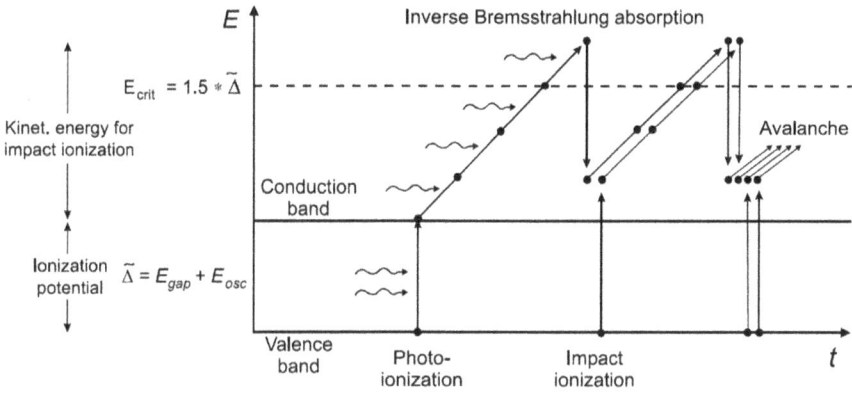

▲ Fig. 3.2. A schematic of photoionization, inverse Bremsstrahlung absorption, and impact ionization in the process of free-electron density growth and plasma generation in numerical simulations. Reprinted from Ref. [1] with permission.

conduction band, and e is the electron charge. The water plasma becomes strongly reflective and absorbing above ρ_{cr}. The critical electron density ρ_{cr} is from 0.984×10^{21} cm^{-3} for $\lambda = 1064$ nm to 3.94×10^{21} cm^{-3} for $\lambda = 532$ nm in numerical simulations. Fig. 3.2 shows the primary processes of laser-induced free-electron growth, including the formation of quasi-free electrons and avalanche ionization. The detailed excitation and growth of free electrons are described below.

Liquid water can be regarded as an amorphous semiconductor, and the excitation energy Δ corresponds to the energy required for electronic transition from the molecular $1b^1$ orbital into an excitation band (band gap 6.5 eV). The band gap energy should be replaced with the effective ionization potential

$$\tilde{\Delta} = \Delta + e^2 F^2 / 4m\omega^2 \tag{3.2},$$

for very short laser pulses [5], where F is the amplitude of the electric laser field, $1/m = 1/m_c + 1/m_v$ is the exciton reduced mass given by the effective masses m_c and m_v of electrons in the conduction and valence bands, respectively. Here we consider the photon energies below the

band gap, thus the photoexcitation mainly proceeds via photoionization (multiphoton ionization or strong-field tunneling) and impact ionization. Once a free electron is produced in the medium, it can absorb photons in the non-resonant "inverse Bremsstrahlung" process via collisions with heavy charged particles. After several inverse Bremsstrahlung adsorption events, the kinetic energy is sufficiently large to produce another free electron through impact ionization, and the critical kinetic energy is $E_{crit} = 1.5\tilde{\Delta}$ for impact ionization. The recurring sequence of inverse Bremsstrahlung absorption and impact ionization leads to the avalanche growth of free electrons, and this process is called "avalanche ionization" or "cascade ionization". There are time constraints for cascade ionization because several consecutive inverse Bremsstrahlung absorption events are needed, thus cascade ionization plays only a minor role in femtosecond breakdown.

The temporal evolution of electron density in the conduction band can be expressed with a generic rate equation [6],

$$d\rho_c/dt = \eta_{photo} + \eta_{casc}\rho_c - \eta_{diff}\rho_c - \eta_{rec}\rho_c^2 \qquad (3.3).$$

The first term represents the production of free electrons via photoionization (multiphoton ionization and strong-field tunneling), the second term represents the contribution of cascade ionization, and the last two terms represent the losses from the diffusion of electrons and electron-hole recombination. Diffusion and recombination do not play a significant role during femtosecond laser pulses. The rate equation can be solved numerically for various laser pulse intensities I using a Runge-Kutta method with adaptive step-size control.

Fig. 3.3(a) and (b) show the temporal evolution of free-electron density ρ_c during the laser pulse at the optical breakdown threshold for 6 ns, 1064 nm pulses and for 100 fs, 800 nm pulses, respectively. The time t is normalized with respect to the laser pulse duration τ_L to facilitate a comparison between the different pulse durations. The contribution of photoionization to the total free-electron density is plotted as a dotted

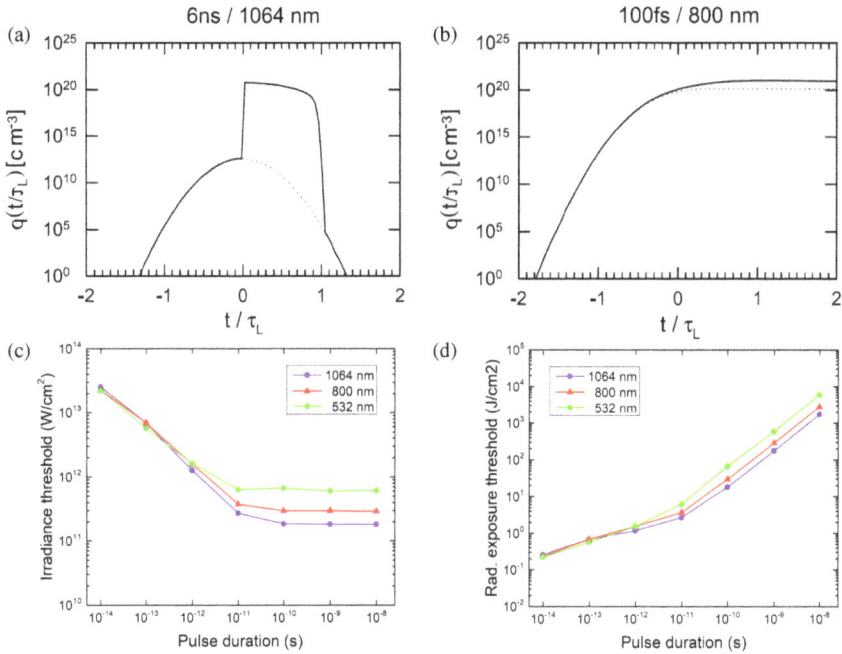

(a) 6ns / 1064 nm

(b) 100fs / 800 nm

(c)

(d)

▲ Fig. 3.3. The electronic excitations and irradiation thresholds for plasma generation in numerical simulations. The evolutions of free-electron density at optical breakdown threshold are shown for (a) 6 ns, 1064 nm pulses and (b) 100 fs, 800 nm laser pulses. The time t is normalized with respect to laser pulse duration τ_L. The contribution of multiphoton ionization to the total free-electron density is plotted as a dotted line. The calculated optical breakdown thresholds ($\rho_{cr} = 10^{21}$ cm^{-3} is used here) of (c) irradiance and (d) radiant exposure as a function of laser pulse duration for various laser wavelengths are also exhibited. Reprinted from Ref. [1] with permission.

line. It is obvious that cascade ionization plays a key role for nanosecond laser pulse excitation. Fig. 3.3(c) shows that a much higher irradiance I_{rate} is necessary for optical breakdown for femtosecond laser pulses, because the generation of free electrons is through multiphoton and strong-field tunneling ionization. Fig. 3.3(d) shows the threshold values of radiant exposure $F_{rate} = I_{rate} \times \tau_L$. The threshold radiant exposure F_{rate} exhibits a weak dependence on pulse duration for $\tau_L < 10$ ps, reflecting the weak recombination and diffusion effects and emphasizing the role of energy absorption. The recombination and diffusion come into effects for longer

pulses, and the threshold irradiance I_{rate} remains approximately constant to provide the seed electrons for cascade ionization.

3.4 Water Dissociation Following Laser Irradiation

Actually, the optical breakdown of liquid water not only involves ionization but also dissociation of water molecules. The ionization of liquid water instantaneously leads to the formation of a hydrated-electron precursor and a cationic hole (H_2O^+), and both species are very reactive. The H_2O^+ undergoes the rapid sub-100 fs proton transfer to a neighboring water molecule to yield the hydronium cation (H_3O^+) and the hydroxyl radical (OH),

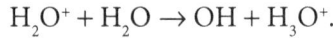

$$H_2O^+ + H_2O \rightarrow OH + H_3O^+.$$

These reactive intermediate radicals may eventually bring about the formation of stable products, e.g., H_2, O_2, and H_2O_2 [7].

With tunable femtosecond soft X-ray pulses, Loh et al. [8] observed the ultrafast dynamics of valence hole created by strong-field tunneling ionization and tracked the proton transfer reaction. The 60 fs, 800 nm laser pulses used have a peak intensity of 2.3×10^{14} W/cm². Fig. 3.4(a) shows that the ionization leads to the emergence of an absorption resonance at 525.9 eV and a redshift of pre-edge absorption, and the new absorption resonance responds to the creation of H_3O^+ and OH. The time-resolved differential absorption spectrum $\Delta A = A(\Delta t) - A(\Delta t < 0)$ demonstrates that the initial species produced by ionization decays with lifetime τ_1 and transforms to other intermediate species with lifetime τ_2 (Fig. 3.4(b)). The two absorption spectra $S_1(E)$ and $S_2(E)$ can be defined accordingly. For example, $S_2(E)$ is calculated by averaging ΔA from 1.5 ps to 5.8 ps, and $S_2(E)$ can be fitted with two Lorentzians with an energy spacing of 0.48 eV. This is consistent with the 0.53 eV spacing of vibrational progression of core-excited gas-phase OH, and $S_2(E)$ can be assigned to the signal of the OH radical. The surface fit of experimental data in Fig. 3.4(b) yields $\tau_1 = 0.18 \pm 0.02$ ps and $\tau_2 = 14.2 \pm 0.4$ ps. The time constants τ_1 and τ_2

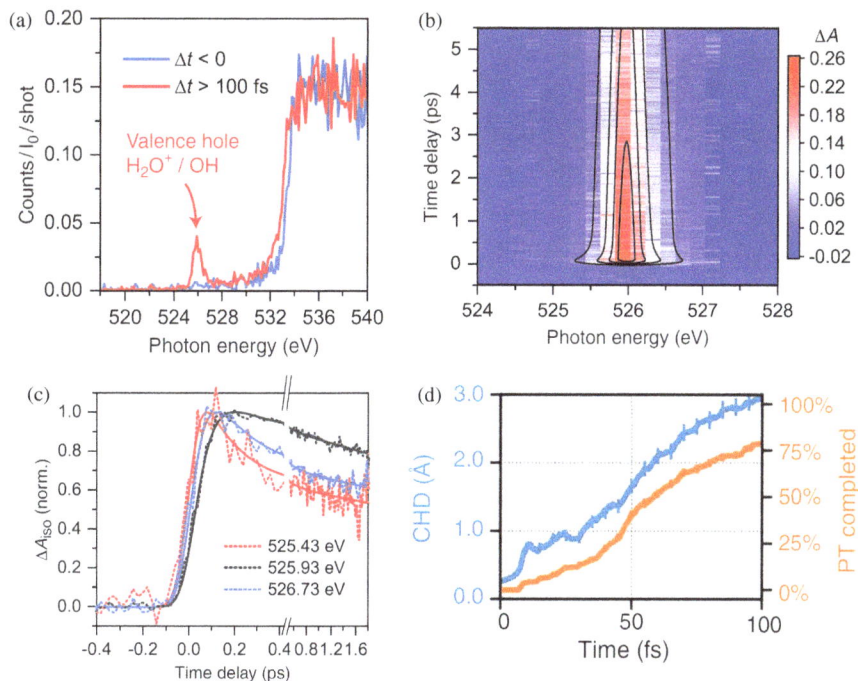

▲ Fig. 3.4.   Transient X-ray absorption spectroscopy accounting for proton transfer reaction and surface-hopping simulations. (a) X-ray absorption for $\Delta t < 0$ and $\Delta t < 100$ fs. (b) Differential absorption ΔA in the valence hole (H_2O^+/OH) region. (c) Polarization-averaged time traces ΔA_{iso} at three X-ray probe energies: 525.43, 525.93, and 526.73 eV. (d) Charge-hole distance and completed proton transfer percentage. Reprinted from Ref. [8] with permission.

correspond to the cooling of a vibrationally hot OH radical and geminate recombination of OH with the hydrated electron respectively. Further insight into the early-time dynamics was obtained with time traces ΔA_{iso} at three selected photon energies: 525.43, 525.93, and 526.73 eV (Fig. 3.4(c)). The signal ΔA_{iso} for 525.93 eV shows a markedly delayed rise relative to those for 525.43 and 526.73 eV, which suggests the existence of an additional ultrafast process. The global fitting of ΔA_{iso} indeed gives a new component with a time constant of 46 fs, and this process reflects the formation of OH radical or the lifetime of H_2O^+ radical.

Nonadiabatic simulations with a surface-hopping approach [9] reveal the microscopic mechanisms of ultrafast structural dynamics. The simulations demonstrate that the cationic ground state is reached within 25 fs when initial ionization is set in the upper 1.5 eV of the valence band. The proton transfer can be characterized by the distance between the charge and hole center, because the charge-hole separation is correlated with the completion of proton transfer (Fig. 3.4(d)). The charge and hole are initially overlapped at the H_2O^+, then the hole stays on the OH moiety while the charge is carried away by the H_3O^+. The timescale of proton transfer is about 60 fs. Moreover, the chemical environment surrounding H_2O^+/OH is strongly correlated with measured X-ray spectral signatures. The negatively charged oxygens are initially pulled toward positive H_2O^+, and this motion leads to the redshift of absorption peak and initiates the proton transfer. After charge-hole separation, the distance between OH and neighboring oxygen increases and the redshift shrinks.

3.5 Nonadiabatic *Ab Initio* Simulation of Nonequilibrium Plasma Generation

The initial atomistic movements and associated energy transfer pathways are hard to track due to the extreme complexity of laser-water interaction and ultrafast timescale. The accurate and complete descriptions of plasma generation processes require real-time nonadiabatic *ab initio* molecular dynamics simulations [10]. Xu *et al.* [11] employed the real-time time-dependent density functional theory (rt-TDDFT) methods to simulate the photoinduced ultrafast dynamics and plasma generation in liquid water. The electron density is propagated in real time through numerical integration of time-dependent Kohn-Sham equations. The liquid water is irradiated with the laser pulses

$$E(t) = E_0 \cos(2\pi\, ct/\lambda)exp[-(t-t_0)^2/2\sigma^2]$$
(3.4),

where the wavelength λ is 800 nm, the width σ is 5 fs, and the center t_0 is 24 fs. The maximum electric field E_0 ranges from 0.7 to 2.4 V/Å, and

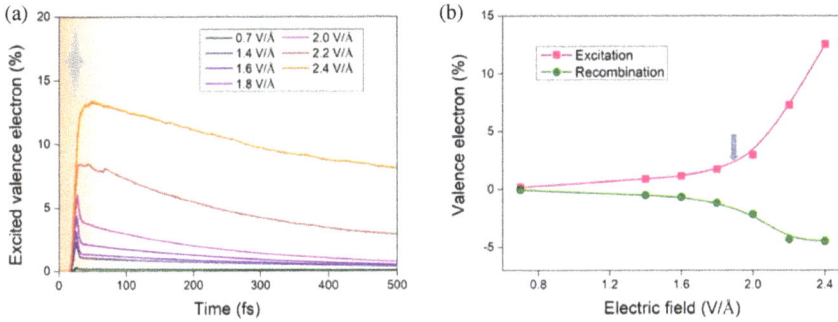

▲ Fig. 3.5. Electronic dynamics of liquid water. (a) Temporal evolutions of excited valence electrons with the electric field of laser pulses. (b) Effective electron excitation and recombination with electric fields of laser pulses. The arrow shows the transition from multiphoton to strong-field ionization processes. Reprinted from Ref. [11] with permission.

the corresponding peak intensities are on the order of 10^{13} W/cm^2. The photoionization (multiphoton ionization and strong-field tunneling) is the dominant mechanism of electronic excitations for ultrashort laser pulses here. Upon photoexcitation, liquid water undergoes successive nonequilibrium evolutions for both ionic and electronic subsystems.

Fig. 3.5(a) shows that the 800 nm laser pulses excite the valence electrons of liquid water, then the excited electrons undergo subsequent relaxation processes via electron-electron and electron-ion interactions. The electronic occupations exhibit the quasi-Fermi-Dirac distributions with electronic subsystems heated to over 10,000 K. The effective electronic excitation can be defined as the ratio of excited valence electrons at $t = 100$ fs. The electronic excitation exhibits a nonlinear relationship with the maximum electric field of laser pulses (Fig. 3.5(b)), and the nonlinearity corresponds to the transition from multiphoton ionization to strong-field tunneling. According to Keldysh's method [5], strong-field tunneling is responsible for ionization at low laser frequencies and large field strengths, while multiphoton ionization become the primary excitation mechanism for optical frequencies and moderate field strengths. The transition from multiphoton to tunneling takes place at field strength $E_0 = 1.9$ V/Å, which is consistent with the estimation of about 1.0–2.0 V/Å for $\lambda = 800$ nm.

The subsequent electron-hole recombination lowers the density of electronic excitation and leads to the cooling of electronic subsystems. Fig. 3.5(b) shows that the electron-hole recombination rate increases with the initial intensity of effective electronic excitations.

The evolution of ionic subsystems is dependent on the time-dependent occupation of electronic subsystems, which determines the nonadiabatic potential energy surfaces. To track the ionic dynamics of photoexcited liquid water, the proton transfer coordinate can be defined as the distance difference between each hydrogen atom and its two nearest oxygen atoms, $\delta = |d_{HO1} - d_{HO2}|$. Fig. 3.6(a) exhibits the temporal evolutions of average δ of all hydrogen atoms in the nonadiabatic simulations. The initial ~1 Å of δ

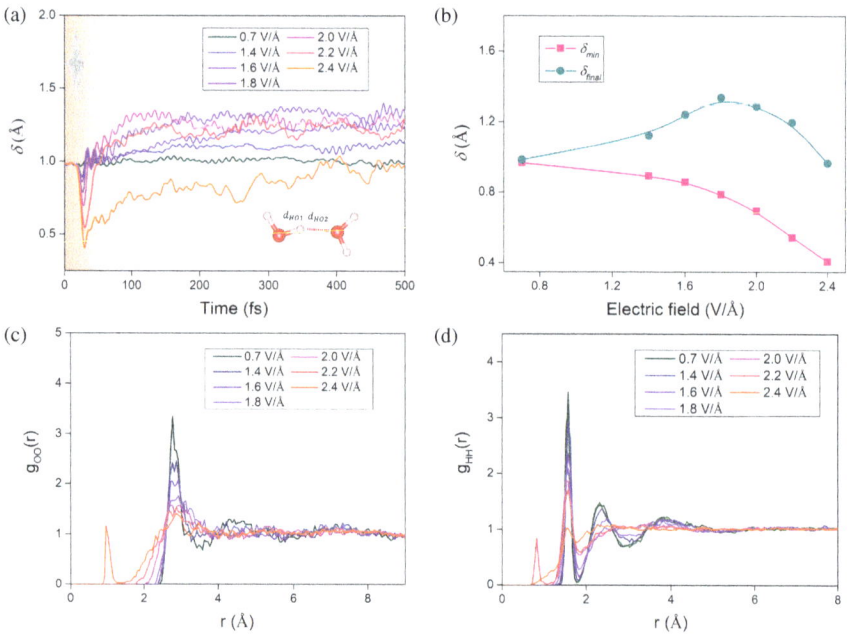

▲ Fig. 3.6. Ionic dynamics of liquid water. (a) Temporal evolutions of average δ of all hydrogen atoms with electric field. (b) Minimum of average δ and the final δ with electric fields of laser pulses. (c) Oxygen-oxygen and (d) hydrogen-hydrogen radial distribution functions with the electric field of laser pulses. The radial distribution functions are calculated with the whole 500 fs-long simulations for each electric field. Reprinted from Ref. [11] with permission.

corresponds to the undisturbed hydrogen bond network of liquid water. The average δ hardly changes for the weak pulses of $E_0 = 0.7$ V/Å, while the average δ undergoes a sudden decrease for stronger laser pulses of $E_0 > 1.4$ V/Å. Fig. 3.6(b) shows that the minimum δ_{min} exhibits a nonlinearity with the electric field. The decrease of δ corresponds to the stretch and break of OH bonds during the laser pulses, reflecting the effects of electric field of laser pulses. After the laser irradiation, the photoexcited water undergoes a fast equilibration of ~100 fs to relax the stretched hydrogen bond network, leading to the increase of δ. Detailed analysis reveals that some protons shuttle between their two nearest oxygen atoms for $E_0 \geq 1.8$ V/Å, and this process leads to dissociation of water molecules for $E_0 \geq 2.2$ V/Å.

Finally, the δ_{final} is reached after the slow equilibrium process of ~300 fs (Fig. 3.6(b)). The δ_{final} increases with maximum electric field under medium-strength laser pulses (1.4 V/Å $\leq E_0 \leq$ 1.8 V/Å), and the increase corresponds to the distortion of hydrogen bond network of liquid water. The δ_{final} begins to decrease for $E_0 = 2.0$ V/Å due to the decrease of average oxygen-oxygen distance (Fig. 3.6(c)) and dissociation of water molecules. In particular, severe water dissociation leads to the generation of free protons amounting to 50% at $E_0 = 2.4$ V/Å (Fig. 3.7(a)). The radial distribution functions demonstrate the disordering of liquid water due to distortion and dissociation of liquid water (Fig. 3.6(c)–(d)). Furthermore, Fig. 3.6(c)–(d) show the formation of transient oxygen and hydrogen molecules following the dissociation of water molecules. In fact, the violent water dissociation gives rise to the generation of various intermediate species in liquid water, e.g., free protons and oxygens, hydroniums, hydroxyls, transient hydrogen molecules and oxygen molecules (Fig. 3.7(b)).

The strong photoexcitation of electronic subsystems is needed to drive the ultrafast structural changes of ionic subsystems within a single ultrafast laser pulse, and avalanche effects hardly contribute to electronic excitations. The laser pulse of $E_0 = 2.4$ V/Å excites ~12% of valence electrons, and severe structural changes are induced with the generation of various transient species (Fig. 3.7(b)). The dissociation of water molecules leads to gap closure and introduces a metallic electronic density of states

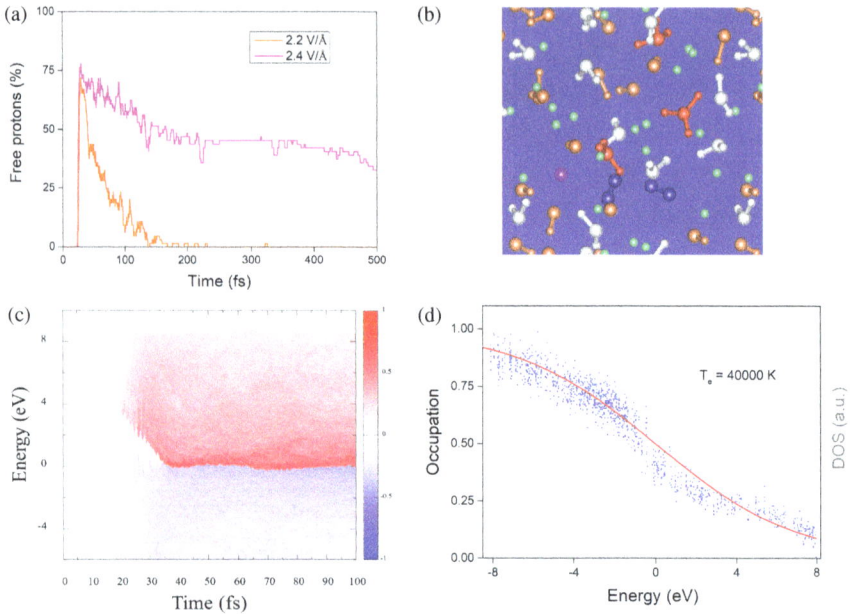

▲ Fig. 3.7. The severe water dissociation and plasma generation in liquid water. (a) The evolution of the ratio of free protons. Free protons are defined as such if the smallest hydrogen-hydrogen distance is greater than 1 Å and the smallest hydrogen-oxygen distance is simultaneously greater than 1.2 Å. (b) The configuration at $t = 500$ fs in nonadiabatic *ab initio* simulations of $E_0 = 2.4$ V/Å. The water molecules, hydroniums, hydroxyls, free protons, oxygen atoms, and oxygen molecules are shown in white, red, orange, green, purple, and dark blue colors, respectively. (c) The evolution of occupation number changes with the laser pulse of 2.4 V/Å. The valence electrons are excited to the conduction band, and the band gap become zero at $t = \sim35$ fs. (d) Electronic occupation at $t = 100$ fs in the simulation with an electric field of 2.4 V/Å. The instantaneous electronic density of states is also shown. Reprinted from Ref. [11] with permission.

(Fig. 3.7(c)), and the instantaneous electronic temperature is ~40,000 K (Fig. 3.7(d)). These confirm the ultrafast plasma generation in liquid water. The threshold for plasma generation is ~7.6 × 10¹³ W/cm² (2.4 V/Å) for the ultrashort laser pulses used in rt-TDDFT simulations. We note that only electron density criteria are considered in previous numerical simulations, and a smaller intensity threshold of ~2 × 10¹³ W/cm² is obtained for the similar 10 fs-long laser pulses in numerical simulations. The obtained water plasma here is the nonequilibrium state of liquid water following the intense photoexcitation. The nonequilibrium water plasma with a density of 1 g/cm³ exhibits a high fraction of free protons (~50%), nonequilibrium electronic and ionic distributions, and a metallic electronic density of states.

The thermodynamic quantities of liquid water are weakly affected by weak laser pulses due to weak photoexcitation (Fig. 3.8(a)–(b)). On the other hand, liquid water can be severely heated and pressurized by strong laser pulses due to intense electronic excitations and violent structural changes. At the threshold E_0 = 2.4 V/Å for plasma generation, the ionic subsystem of liquid water is heated by ~3,500 K and the system pressure is simultaneously increased by ~100 kbar within 500 fs. The change of

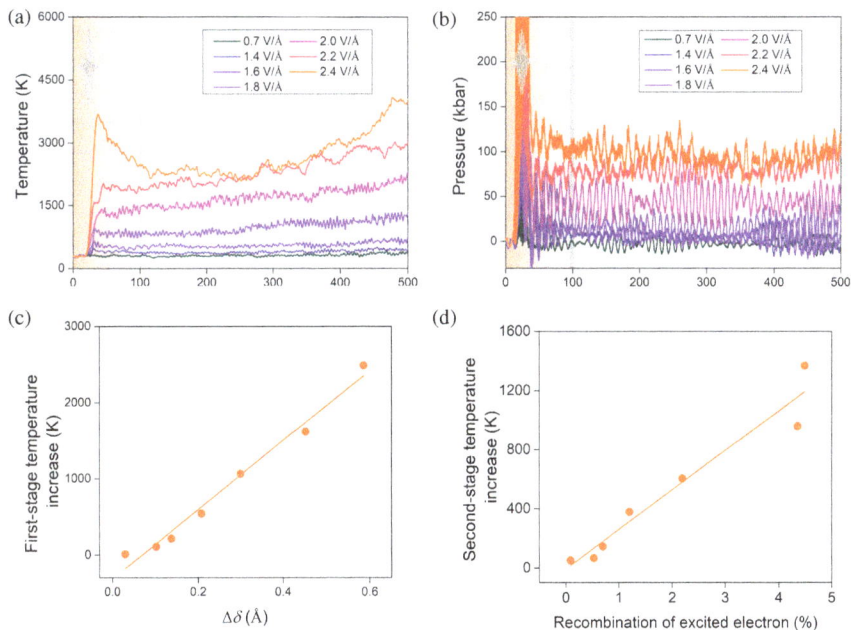

▲ Fig. 3.8. Photoinduced thermodynamics of liquid water. (a) Temporal evolutions of ionic temperatures with the electric field of laser pulses. The light envelope is shown in the inset. The vertical line marks the separation of the first and second stages. (b) The temporal evolutions of system pressures of photoexcited liquid water. (c) Decrease of average δ vs the first-stage ionic temperature increase. (d) Electron recombination vs the second-stage temperature increase of ionic subsystems. Reprinted from Ref. [11] with permission.

ionic temperature is proportional to that of system pressure. The ionic temperature undergoes the two-step heating process separated at $t = 100$ fs. The first stage is coincident with the envelope of the laser pulses ($t < 50$ fs), and the second-stage heating takes place during the relaxation process of photoexcited liquid water. The decrease of δ during laser pulses is strongly correlated with the first-stage increase of ionic temperature (Fig. 3.8(c)), while the electron-hole recombination leads to the second-stage increase of ionic temperature (Fig. 3.8(d)). The contributions from the first stage are larger than those of the second stage within the 500 fs-long simulations. The evolution of system pressure also exhibits the two-step process (Fig. 3.8(b)). The electron excitation greatly enhances the internal pressure of liquid water and dominates the pressure increase in the first stage, while the relaxation processes lead to the slight decrease of pressure in the second stage.

3.6 Conclusion

In this chapter, we reviewed the observations of laser-induced plasma generation in liquid water and discussed the underlying mechanisms of nonequilibrium plasma generation in numerical and quantum mechanical simulations. The severe optical breakdown and water dissociation give rise to plasma generation in liquid water. In numerical simulations, breakdown threshold is defined as a critical free-electron density ρ_{cr} or the irradiance required to produce that free-electron density. Avalanche ionization dominates the electronic excitations for picosecond and nanosecond laser pulses, while photoionization (multiphoton effects and strong-field tunneling) is the primary mechanism of electronic excitations for ultrashort laser pulses. At the same time, the optical breakdown of liquid water leads to both ionization and dissociation of water molecules. Therefore, rt-TDDFT simulations are necessary for accurate and complete descriptions of the ultrafast generation of nonequilibrium water plasma. The separative evolutions of electronic and ionic subsystems and the energy transfer pathways can be explicitly monitored in rt-TDDFT simulations.

Femtosecond laser pulses (~10 fs) stretch water molecules during laser irradiation, and electronic excitation may lead to proton transfers after laser pulses. Intense laser pulses (~7.6 × 10^{13} W/cm^2) drastically excite liquid water, giving rise to severe water dissociation and the generation of water plasma. The nonequilibrium water plasma was shown to exhibit separate ionic and electronic temperatures, metallic electronic density of states, and highly dissociated atomic configurations.

References

1. Vogel A., Noack J., Hüttman G. & Paltauf G. (2005). Mechanisms of femtosecond laser nanosurgery of cells and tissues, Applied Physics B, 81, pp. 1015–1047.
2. Schaffer C. B., Nishimura N., Glezer E. N., Kim A. M.-T. & Mazur E. (2002). Dynamics of femtosecond laser-induced breakdown in water from femtoseconds to microseconds, Optics Express, 10, pp. 196–203.
3. Klein A. L., Bouwhuis W., Visser C. W., et al. (2015). Drop shaping by laser-pulse impact, Physical Review Applied, 3, pp. 044018.
4. Stan C. A., Milathianaki D., Laksmono H., et al. (2016). Liquid explosions induced by X-ray laser pulses, Nature Physics, 12, pp. 966–971.
5. Keldysh L. V. (1965). Ionization in the field of a strong electromagnetic wave, Soviet Physics Journal of Experimental and Theoretical Physics, 20, pp. 1307.
6. Noack J. & Vogel A. (1999). Laser-induced plasma formation in water at nanosecond to femtosecond time scales: calculation of thresholds, absorption coefficients, and energy density, IEEE Journal of Quantum Electronics, 35, pp. 1156.
7. Chin S. L. & Lagacé S. (1996). Generation of H_2, O_2, and H_2O_2 from water by the use of intense femtosecond laser pulses and the possibility of laser sterilization, Applied Optics, 35, pp. 907–911.
8. Loh Z.-H., Doumy G., Arnold C., et al. (2020). Observation of the fastest chemical processes in the radiolysis of water, Science, 367, pp. 179–182.
9. Tully J. C. (1990). Molecular dynamics with electronic transitions, Journal of Chemical Physics, 93, pp. 1061–1071.

10. Lian C., Guan M., Hu S., Zhang J. & Meng S. (2018). Photoexcitation in solids: First-principles quantum simulations by real-time TDDFT, Advanced Theory Simulation, 1, pp. 1800055.
11. Xu J., Chen D. & Meng S. (2021). Probing laser-induced plasma generation in liquid water, Journal of the American Chemical Society, 143, pp. 10382–10388.

© 2024 World Scientific Publishing Company
https://doi.org/10.1142/9789811265648_0004

Chapter 4

Plasma-based Terahertz Wave Generation in Liquids

Anton Tcypkin, Evgenia Ponomareva,[††] Azat Ismagilov

ITMO University, Russia

4.1 Introduction

Soon after the invention of pulsed lasers, which make it possible to generate high-power radiation, researchers actively began to study the processes of nonresonant interaction of light with dielectric media. In the 1990s, for these types of lasers, hot research areas were local modifications of materials and applications in medical treatment. Several scientific groups studied the interaction of pulsed radiation with the liquid state of matter — one of the main components of the biological sphere of our planet.

Initially, studies were mainly focused on ultrashort (<1 ps) pulses [1, 2]. These studies aimed to estimate the threshold values of laser system parameters for optical breakdown excitation, which is complete or partial ionization of the medium. The ionization of matter occurs due to absorption of laser radiation energy and results in the formation of plasma, an electrically quasi-neutral "gas" of charged particles.

[††] This affiliation is not the current place of employment of E. Ponomareva, but is the place where the results described in this chapter were obtained.

The research of the last decades has shown that plasma excitation in a liquid medium leads to the generation of strong electromagnetic fields: from the far-infrared (IR) to ultraviolet or X-ray ranges of the spectrum [3–5]. Liquid as a target for such radiation sources gives greater flexibility due to its self-replacement after it is destroyed by high-intensity radiation. In 2017, a group of scientists demonstrated an alternative use of optical breakdown in a liquid as a source of pulsed terahertz (THz) radiation, which opens prospects for further investigation [6].

In this chapter, we discuss in detail the generation of pulsed THz radiation during plasma formation in various liquids. To do this, we present the most interesting features of liquid-based THz sources; describe a theoretical approach to analyze the mechanism of THz wave generation during plasma formation in a dielectric medium; and discuss a technique for enhancing the generation of THz fields during double-pulse excitation.

4.2 Plasma-based Broadband THz Pulse Generation System with Plane-Parallel Liquid Jets

The first research results revealed high potential of new broadband THz sources based on filamentation in liquids. That fact justified the systematic study of experimental conditions to determine the efficiency of optical-to-THz conversion. Such conditions primarily include the parameters of a laser system and characteristics of a liquid medium used. Further in the text, we consider the features of the optical pumping system and the most important results of its application.

Fig. 4.1(a) shows a typical laser experimental setup to generate broadband THz radiation in plane-parallel jets of liquids. For example, a femtosecond laser system with a central wavelength of 800 nm, a pulse duration of 35 fs to 700 fs, and a pulse energy of up to a few mJ can be used as an optical pump [7]. The pulse duration is decreased or increased by changing the distance between the compressor gratings to form a positive or negative chirp, respectively. The pump pulse is divided into two parts by a beam splitter (BS) for the probe and pump pulses. The pump radiation

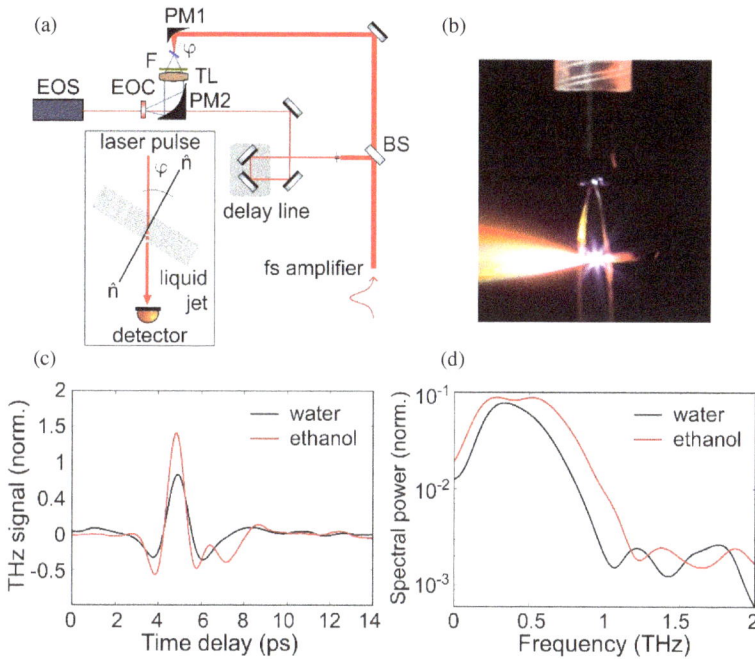

▲ Fig. 4.1.   Experimental setup for THz generation in plane-parallel liquid jets. (a) Experimental scheme for energy and spectrum measurements of the generated THz radiation
(the inset shows the optical incident angle φ). (b) Image of laser excitation of the liquid jet.
Water moisture plum scatter the laser beam. (c) Typical THz temporal signals and (d) spectra emitted from the jets of water and ethanol. Reprinted from Ref. [7] with permission.

is focused by a parabolic mirror with a focal length of 5 cm (PM1) and
a caustic of 100 μm onto a plane-parallel liquid jet (thickness 100 μm,
150 μm, or 270 μm). The interaction of intense radiation with a liquid
medium leads to its ionization, and because of plasma formation, pulsed
THz radiation is also generated. To form jets of different thicknesses,
special nozzles of certain sizes are selected. The probe pulse passes through
the delay line and is detected by the electro-optical system (EOS). The use
of a ZnTe crystal with a thickness of 1 mm allows detecting a signal up to
3 THz.

The energy of THz radiation depends on the incident angle between
the pump radiation and the normal to the jet surface; therefore, such

experimental setups use systems that allow changing the angle of incidence φ (Fig. 4.1(b), where φ is the normal to the jet surface). The optimal incident angle is determined by calculating the transmission of p-polarized pump radiation at the air/liquid interface and the direction of the dipole. For example, the corresponding value φ for water is 65°, and for ethanol it is 60°. Fig. 4.1(c) and (d) show the temporal signals and corresponding spectra of THz radiation generated by filamentation in a 150 μm-thick jet of water or ethanol with pump laser pulse duration of 400 fs and an excitation energy of 600 μJ, respectively.

The formation of a plane-parallel jet is a challenging task, technically realized several decades ago, for example, for the dye lasers. However, the final integration of this system as an element of the experimental setup requires additional efforts. A general view of the system for the liquid jet formation is presented in Fig. 4.2(a). The system consists of (1) a pump, (2) a power supply, (3) a damping tank, and (4) a nozzle with a liquid collection reservoir, tubes, and fittings. The reservoir is filled with liquid, then the liquid is driven by a pump (1) powered by a standard laboratory power supply (2). The power supply with the current regulator allows for the control of the pump speed and, thereby, regulates the flow rate of the liquid. The damping tank (3) is necessary to ensure the uniformity of the flow and the liquid jet width.

To form a plane-parallel jet, this system uses nozzles, which can be a plastic or metal tube with two blades on both inner sides; one end of the tube is compressed [8]. The nozzle is installed in a rotating mount and located on a three-coordinate translation stage, which allows controlling the position of the liquid jet accurately in space and the angle of its rotation relative to the pump radiation. Changes in the geometric dimensions of the nozzle and flow rate provide a jet with a width in the range of 100–800 μm and a thickness in the range of 50–400 μm. A typical nozzle drawing is shown in Fig. 4.2(b). The dimensions marked with the dotted line in the figure are responsible for the jet thickness. Due to 3D printing technologies, the manufacture of nozzles is not a difficult task. Depending on the type of solvent used to generate THz radiation, the nozzles can be made of either plastic or stainless steel.

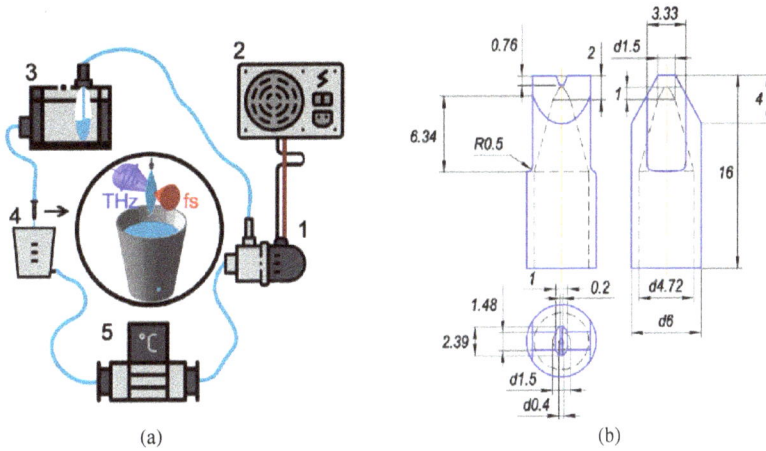

▲ Fig. 4.2. (a) The liquid jet formation system, which is built into the optical setup: (1) liquid pump, (2) power supply, (3) damping tank, (4) nozzle with liquid collection reservoir, and (5) temperature control system. Inset: 3D geometry of the liquid jet with focused pump, IR radiation, and generated THz radiation. (b) The schematic drawing of a nozzle for producing the flat plane-parallel liquid jet. Reprinted from Ref. [9] with permission.

After leaving the nozzle, the liquid enters the collection reservoir, thereby forming a closed cycle. This allows significant reduction of the volume of liquids required for continuous work with the system. Also, this system can be easily modified. For example, to carry out studies related to the temperature of liquids, a temperature control system (5) can be connected to the system.

With the system described, several experiments were conducted to obtain the key dependences characterizing the new source of pulsed THz radiation. For example, the dependence on the pump pulse duration reveals a clear optimum in liquids [6], while the same dependence for generation in gases has the form of a power-law decay of energy with increasing duration. The clear contrast in the measurement results for the two types of matter can be interpreted through the differences in their ionization mechanism. The presence of an optimal value for the duration of the pump pulse for THz generation in liquids was found independently in various works [6, 47]. Fig. 4.3 reveals the correlation between the change in the thickness of the liquid jet and the optimal pulse duration. It is assumed that this feature is related to the relationship between the thickness of the medium and the spatial size of the pump pulse. The optimal value is observed when the medium is "completely" filled with radiation, i.e., when the spatial size of the pump pulse becomes commensurate with its thickness. The subsequent attenuation of the THz radiation energy occurs due to a decrease in the pump intensity with an increase in the pulse duration. Moreover, this dependence is symmetrical with respect to the change in the sign of the phase modulation coefficient (or chirp).

In addition, Fig. 4.3 shows the dependence of the generated THz radiation energy on the pump radiation energy. This dependence is one of the key ones for identifying possible thresholds or limits for the conversion of optical pumping into THz waves in a certain medium. Our measurements have shown that in the case of liquid targets, the dependence of the THz field energy on the pump energy has a quasi-quadratic character (Fig. 4.4). The patterns obtained require theoretical justification, which is given in the following section.

In the above experiments, distilled water was used as the active medium. Water is a polar liquid, therefore like any other polar substance, it has intense absorption bands in the spectrum at THz frequencies. Water is simple to use, but other liquid media with more promising characteristics

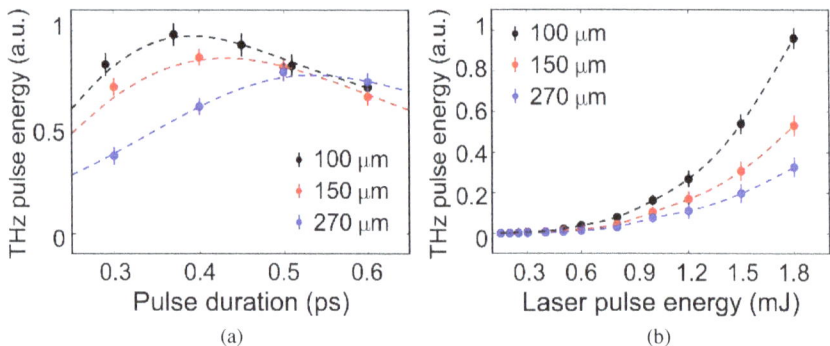

▲ Fig. 4.3. Dependence of the normalized energy of the THz field during filamentation in plane-parallel water jets of different thicknesses on the duration of (a) the pump pulse and (b) the pump energy. Reprinted from Ref. [7] with permission.

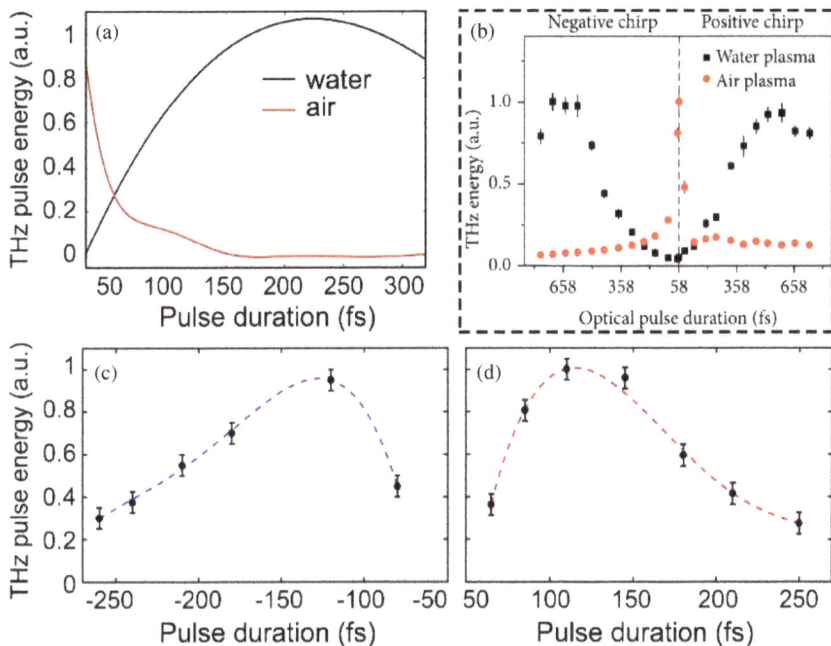

▲ Fig. 4.4. (a) The numerical simulation result of the THz energy dependence on the pump pulse duration for a model medium with parameters of water (black) and gas (red) and (b) a corresponding comparison with experimental data. Numerically simulated THz energy dependence on the pump pulse duration for a (c) negative and (d) positive chirp compared to experimental dots. Reprinted from Ref. [6] with permission.

should be considered to increase the efficiency of optical signal conversion to THz.

First, it is necessary to indicate the influence of linear properties of the liquid on the efficiency of broadband THz source. For example, the change in concentration of various substances in water should modify the absorption spectrum of a solution in the THz frequency range. Thus, our first attempt was to increase the kH of water (the concentration of sodium bicarbonate $NaHCO_3$ in it, or so-called 'hardness'). As seen in Fig. 4.5(a), however, when the water hardness rises, the energy of the generated THz radiation decreases [10].

Furthermore, Fig. 4.5(b) provides the results of experiments on the efficiency of THz wave generation via change in water acidity. The pH value

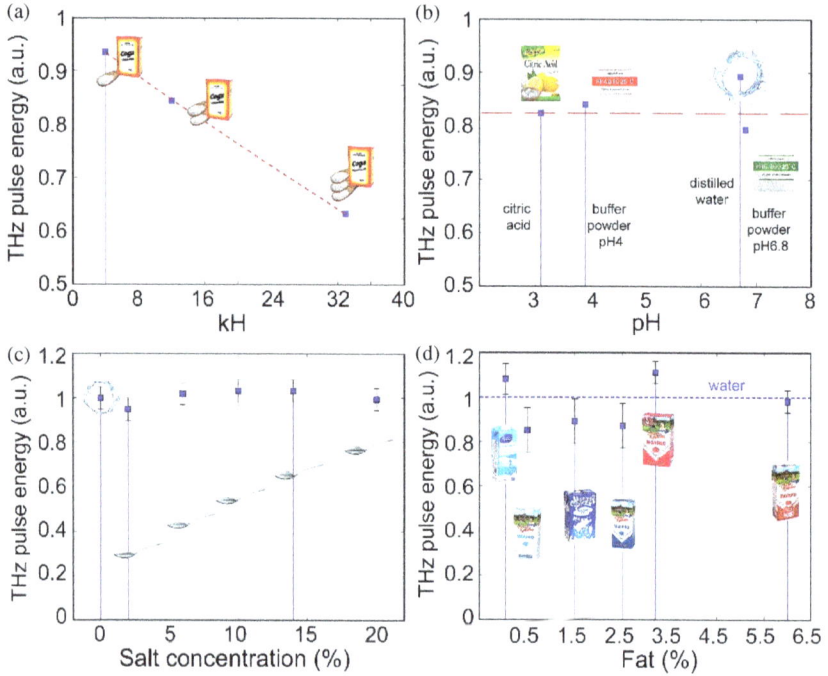

▲ Fig. 4.5.   The energy of the THz radiation pulse during single-color pumping depending on (a) hardness (kH value), (b) acidity (pH value), (c) salt concentration of an aqueous solution, and (d) milk with different percentages of fat. The optical pulse energy and duration was 1 mJ and 200 fs, correspondingly. Reprinted from Ref. [10] with permission.

is lowered with citric acid. The measurements reveal no direct dependence, which leads to the conclusion that the change in hydrogen activity does not affect the generation of radiation in the THz range. Additionally, an attempt to change the efficiency of generating THz radiation is made by varying the salt concentration. However, this has negligible effect on the THz generation efficiency as seen in Fig. 4.5(c).

Finally, the efficiency of optical-to-THz conversion was also measured in milk with different percentages of fat. Based on the results of [11], it was assumed that an increase in the content of milk fat should lead to an increase in the energy of THz waves since the absorption of liquid in the

THz frequency range decreases. However, the results obtained (Fig. 4.5(d)) do not reveal the expected relationship.

Since changes in the chemical properties of water lead to an ambiguous increase in the generation of THz waves, the question about the potential of using other liquids arises. As mentioned above, the dominant mechanism responsible for the generation of THz radiation is plasma formation. In this regard, the choice of liquids is due to the search for the optimal ratio between the molecular density of the medium and the minimum potential of its ionization for the efficient generation of quasi-free electrons. Linear absorption in the THz frequency range is one of the main factors, so nonpolar liquids with a negligibly small absorption coefficient are needed. Table 4.1 lists the parameters of the liquids studied.

In Fig. 4.6(a) the energy of generated THz radiation upon single-color excitation of the studied liquids is compared with the THz energy in the case of two-color air pumping. For the two-color air filamentation, the pulse duration is 35 fs (for maximum optical-to-THz conversion efficiency [30]), and when using a liquid jet with a thickness of 100 µm, the pulse duration is 250 fs.

After normalizing the data to the value obtained using a water jet, it is noticeable that α-pinene is the most promising as its output THz energy is 2.2 times higher. α-pinene is a nonpolar liquid with a negligible linear

▼ Table 4.1. Parameters of the liquids.

Liquid	$n_2 \cdot 10^{-16}$, cm²/W	Ip, eV	P, kg/m³	αTHz, cm⁻¹
Water	1.9 [12]	9 [16]/10.9 [17]	997 [22]	200 [26]
Heavy water	2.4 [13]	12.6 [18]	1104 [22]	100–200 [27]
Ethanol	5 [14]	9.7 [17]	787 [22]	60 [28]
Ethylene glycol	—	10.16 [19]	1123 [23]	~40 [26]
Isopropanol	—	10.15 [20]	781 [24]	~40 [29]
α-pinene	15 [15]	8.04 [21]	859 [25]	~1 (nonpolar)

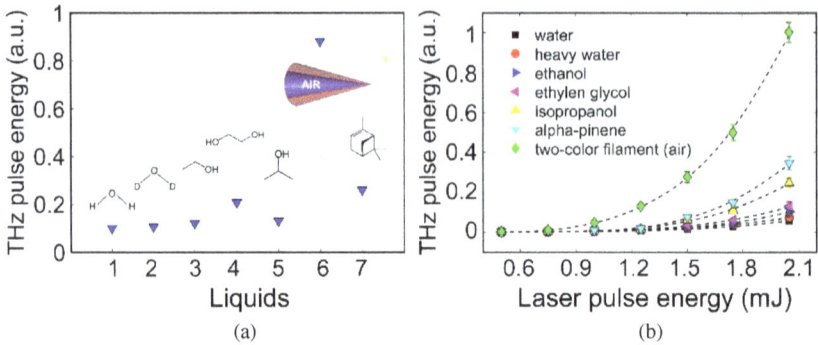

▲ Fig. 4.6.   (a) Comparison of various liquids (1 – water, 2 – heavy water, 3 – ethanol, 4 – ethylene glycol, 5 – isopropanol, 7 – α-pinene) for the generation efficiency of THz radiation from one-color laser filament and 6 – for two-color filamentation in air. (b) The dependence of THz radiation energy during laser filamentation in jets of various liquids on the laser pump energy compared to the result of two-color air filamentation. Reprinted from Ref. [10] with permission.

absorption coefficient in the THz frequency range. Fig. 4.6(b) demonstrates the dependence of the THz radiation energy on the pump pulse energy for the same liquids. The dependence has a quasi-quadratic character, and the energy ratio is preserved for all the values of the pump pulse energy for each liquid medium under consideration.

4.2.1 Theoretical Description of Broadband THz Radiation Emission During Plasma Formation in Liquids

Literature devoted to the ultrashort pulse dynamics in a medium with compound nonlinearity analyzes the evolution of the complex envelope of a laser pulse. However, the complex envelope analysis ceases to be justified during the propagation of a near-IR range laser pulse through the medium, which results in the generation of such broadband radiation that it overlaps the far-IR range as well. In those cases, the dynamics of the radiation field itself are studied, rather than its envelope [31–33]. For example, in [33], to describe the generation of THz waves in liquid nitrogen, a field approach was applied to a photocurrent model. In addition, it was shown in [7] that the field equations describe well the generation of THz waves in plasma induced by intense sub-picosecond Ti:Sa laser pulses in water. To analyze the obtained experimental data, the model [34] is used; it is described by the following equations:

$$\left\{ \begin{array}{l} \dfrac{\partial E}{\partial z} + \Gamma_0 E - a\dfrac{\partial^3 E}{\partial \tau^3} + gE^2\dfrac{\partial E}{\partial \tau} + \dfrac{2\pi}{cn_0}j = 0 \\[3mm] \dfrac{\partial j}{\partial \tau} + \dfrac{j}{\tau_c} = \beta\rho E^3 \\[3mm] \dfrac{\partial \rho}{\partial \tau} + \dfrac{\rho}{\tau_p} = \alpha E^2 \end{array} \right. \qquad (4.1)$$

where n_0 and a are introduced to describe the normal dispersion; $n(\omega) = n_0 + ca\omega^2$ in liquids (such a simple dispersion law is usually valid for most of the radiation spectrum, which in many cases even under overbroadening conditions remains in the normal group dispersion region); Γ_0 is the empirical coefficient, which characterizes the dependence of absorption of the medium, $k(\omega) = c\Gamma_0/\omega$, on frequency ω; g determines the Kerr inertialess nonlinearity and is related to the coefficient of the nonlinear refractive index of the medium by the $g = 2n_2/c$ relation [35]; ρ and j are the population of highly excited levels and current density of quasi-free carriers, respectively; α and β are empirical coefficients [34] determining changes in the population of highly excited levels and current density of quasi-free carriers, respectively; τ_c and τ_p are the times of collisional relaxation and relaxation of highly excited levels, respectively; z is the pulse propagation direction; and $\tau = t - zn_0/c$ is the time in a moving reference frame, where t is time and c is the speed of light in vacuum. Since thin jets (~100 μm) are usually used in experiments on the generation of THz radiation in liquids (for mitigating the effect of THz waves absorption), diffraction is neglected in the mathematical model.

The first equation of Eq. (4.1) describes the evolution of the electric field considering the linear dispersion of the refractive index, the low-inertia cubic nonlinear response, and the plasma nonlinearity. The second equation characterizes the dynamics of the current of quasi-free electrons under the

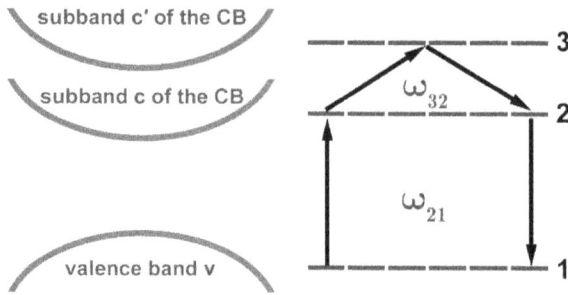

▲ Fig. 4.7.   Three-band energy model of an isotropic dielectric medium (v is the valence band, c and c' are the subbands of the conduction band; 1 is the ground state, 2 and 3 are the excited states of the valence electron).

influence of the field of an ultrashort pulse. The cubic dependence for the field results from time evolution of the current density proportional to the electric field, E, and the number of quasi-free electrons, whose inertial transition from excited states is determined by E^2. The third equation characterizes the change in the population of excited energy states, the transition to which is allowed from the ground state and is determined by the order of E^2. According to the equations described above, ionization is a stepwise process (Fig. 4.7): initially, the electric field excites high-energy states of the molecule, then the transition of electrons to a quasi-free state occurs. In the case of the quantum mechanical model, the authors of [34] study the short duration of optical pulses and the overbroadening of the spectrum during propagation in a medium. Energy transitions in a molecule are considered nonresonant (the resonant multiphoton process is not taken into account due to the wide radiation spectrum) and, therefore, the plasma nonlinearity of the medium response to pulsed radiation is described to be proportional to the fifth order in the field.

It is assumed that the radiation field at $z = 0$ is created by a phase-modulated (chirped) Gaussian pulse [36]:

$$E(\tau) = E_0 exp\left(-\frac{\tau^2}{\tau_0^2}\right) sin\left(\omega_0\tau + \frac{A}{\tau_0}(\omega_0\tau)^2\right), \qquad (4.2)$$

where E_0 is the maximum pulse amplitude at the input surface, ω_0 is the central frequency of radiation at central wavelength $\lambda_0 = 800$ nm, A is the phase modulation coefficient chosen so that the spectral width of the phase-modulated pulse corresponds to the spectral width for a spectrally limited pulse with a duration of 35 fs, and τ_0 is the pulse duration. To take into account the influence of the nonlinear properties of liquids on the efficiency of optical-to-THz conversion, we first neglect linear absorption in the medium.

Eq. (4.1) could be easily reduced to one equation. Additionally, it will be convenient to introduce dimensionless parameters:

$$\tilde{E} = E / E_0, \tilde{z} = za\langle\omega\rangle^3, \tilde{\tau} = \tau\langle\omega\rangle.$$

Thus, we obtain:

$$\frac{\partial \tilde{E}}{\partial \tilde{z}} - \frac{\partial^3 \tilde{E}}{\partial \tilde{\tau}^3} + \tilde{g}\tilde{E}^2 \frac{\partial \tilde{E}}{\partial \tilde{\tau}} + \tilde{g}_P \int_{-\infty}^{\tilde{\tau}} \tilde{E}^3 e^{-\frac{(\tilde{\tau}-\tilde{\tau}')}{\langle\omega\rangle\tau_c}} d\tau' \int_{-\infty}^{\tau'} \tilde{E}^2 e^{-\frac{(\tau'-\tau'')}{\langle\omega\rangle\tau_p}} d\tau'' = 0 \quad (4.3)$$

Here, $\tilde{g} = g\frac{E_0^2}{a\langle\omega\rangle^2}$ and $\tilde{g}_P = g_P\frac{E_0^4}{a\langle\omega\rangle^3}$, where $g_p = \frac{2\pi}{cn_0}(\alpha \cdot \beta)_{liq}$. Fig. 4.8 compares the results of the experiment measurements [7] and numerical calculation [37] of the dependence of the energy of the THz radiation pulse on the energy of the pump pulse when it was generated in a plane-parallel

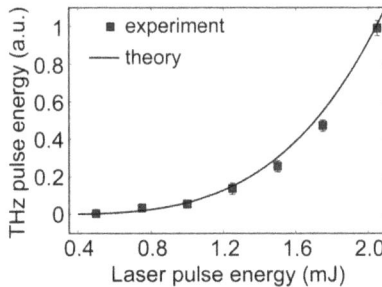

▲ Fig. 4.8. Experimental and simulated dependence of the THz radiation pulse energy on the pump pulse energy generated in a plane-parallel water jet. Reprinted from Ref. [37] with permission.

150 μm-thick water jet. Pump radiation parameters are central wavelength 800 nm and pulse duration τ_{pulse} = 400 fs. The following parameters for a liquid medium are used: a = $3.6 \times 10^{-44}s^3$/cm, g = 1.4 10^{-24}cm × s/W, g_p = $4.5 \times 10^{10}cm^3/(W^2 \cdot s^2)$, τ_c = 2 fs [1] and τ_p = 150 fs [38]. Substantial agreement between the experimental data and numerical simulation results was achieved.

The above-normalized coefficients of Eq. (4.3) make it possible to determine the degree of influence of each physical process on the change in the pulse during its propagation, taking into account the individual contributions of Kerr and plasma nonlinearities to the generation of THz radiation.

Fig. 4.9 shows the THz signal generated by considering individual nonlinearities, as well as their total contribution.

As seen from Fig. 4.9, the contribution of the third-order nonlinearity to the generation of THz radiation in liquids is insignificant with respect to the contribution of plasma nonlinearity. Accordingly, one of the main mechanisms for generating THz radiation is the free electron dynamics in the plasma channel. Moreover, when the plasma and Kerr nonlinearities are taken into account mutually, the THz signal decreases due to the redistribution of energy between the third- and fifth-order nonlinearity mechanisms.

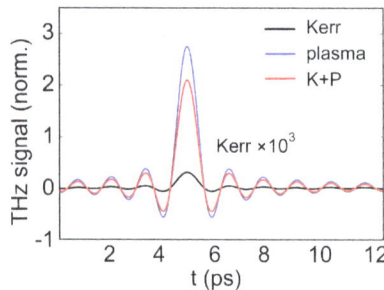

▲ Fig. 4.9. Temporal form of the generated THz pulse, considering the contribution of only the Kerr (black) or plasma nonlinearities (blue), as well as their total contribution (red). Reprinted from Ref. [37] with permission.

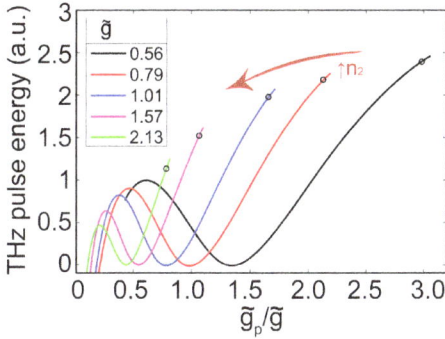

▲ Fig. 4.10.   Dependences of the pulse energy of the generated THz radiation with an increase in the ratio $\tilde{g}_p/\tilde{g}$ at fixed values of the normalized Kerr nonlinearity coefficient $\tilde{g}$. Reprinted from Ref. [37] with permission.

Hence, it becomes necessary to study the effect of the $\tilde{g}_p/\tilde{g}$ (Eq. (4.3)) ratio on the efficiency of THz generation during filamentation in a liquid. Fig. 4.10 shows the results of numerical simulation of the pulse energy generated by THz radiation with an increase in the plasma nonlinearity contribution at various fixed values of the third-order nonlinearity contribution.

For each fixed $\tilde{g}$ value, there is a typical curve, which can be conditionally divided into two characteristic regimes for generation of THz radiation. The first corresponds to the case where the energy of the THz radiation pulse under the dominance of the third-order effects grows and then decreases because of the destructive mutual influence of the Kerr and plasma nonlinearities. The second regime demonstrates a sharp increase in the energy of the generated THz pulse and starts after a characteristic minimum, which can be associated with the ionization threshold of the medium. In this case, the plasma nonlinearity contribution dominates over the third-order nonlinearity contribution.

Moreover, with an increase in the contribution of the Kerr nonlinearity, the described minimum (transition between regimes) shifts towards lower values of the ratio of normalized coefficients. The ratio of the normalized coefficients can be expressed in terms of $\tilde{g}_p/\tilde{g} = g_p E_0^2/g\langle\omega\rangle$, which shows that it is proportional to the pump pulse intensity.

The arrow and dots in Fig. 4.10 show a decrease in the THz radiation pulse energy with an increase in the normalized Kerr nonlinearity coefficient for a fixed plasma effect in the strong ionization regime. This is due to the pump pulse energy redistribution to the third-order mechanisms.

We note that the model described in this section was applied to all numerical experiments and theoretical analysis, which is presented below.

The experimentally discovered quasi-quadratic nature of the pulse energy dependence on the pump radiation energy has no simple theoretical interpretation. In accordance with Eq. (4.1), the Kerr nonlinearity has cubic dependence on the field while the plasma nonlinearity has fifth-order dependence. If we carry out multiplication of Eq. (4.1) by field E, we obtain that the derivative of the square with respect to the field is proportional to the nonlinear term with the fourth and sixth orders in the field. Thus, the square energy dependence of the pump pulse occurs due to the Kerr nonlinearity, and the cubic dependence results from the nonlinearity of the self-induced plasma. Taking into account all the approximations, these estimates demonstrate an almost quadratic and cubic dependence of the THz radiation pulse energy on the pump intensity, which is in good agreement with the experimental results. Moreover, the quasi-free electron density dynamics with a square dependence in the field and in the emerging plasma current proportional to the pump radiation field contribute to the quasi-quadratic nature of the dependence of the THz radiation pulse energy generated during laser filamentation in liquid jets on the incident IR radiation pulse energy.

Another feature, which was also reproduced by using the proposed theoretical model, was the correlation between the optimal pulse duration and the jet thickness (Fig. 4.11(a)). Fig. 4.11(b) compares the experimental and simulation results of the dependence of the THz pulse energy on the duration of a pump pulse with an energy of 600 µJ in plane-parallel water jets of various thicknesses (100 µm (black), 150 µm (red), and 270 µm (blue)). With the help of simulation, the THz radiation energy values for the missing experimental data are restored, additionally confirming the assumption that the spatial size of the pump pulse must correspond

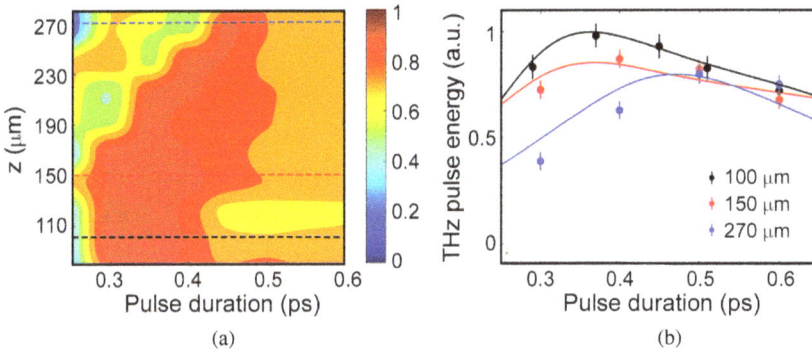

▲ Fig. 4.11. (a) Numerical simulation of the THz energy dependence on propagation distance in plane-parallel water jets for various pump durations. (b) The comparison of experiment results (dots) and simulation dependences (lines) of the THz energy on the laser pulse duration for the pump energy of 600 μJ in plane-parallel water jet of different thickness (100 μm (black), 150 μm (red) and 270 μm (blue)). Reprinted from Ref. [7] with permission.

to the thickness of the medium. Furthermore, the simulation takes into account the linear absorption of water, which confirms the experimentally demonstrated decay of the THz field energy with increasing jet thickness. Thus, the absorption of the medium in the THz range is one of the key parameters when choosing a liquid.

According to the analysis performed, the model used allows us to describe the processes of generation of THz radiation during plasma formation in liquids. The data obtained revealed that the efficiency of optical-to-THz conversion during filamentation of sub-picosecond pulses in liquid jets depends on the linear absorption of liquids in the THz frequency range, as well as on their plasma nonlinearity coefficient, which is proportional to the molecular density of the liquid and inversely proportional to its ionization energy.

Based on the results demonstrated above, we can conclude that liquid-based THz radiation sources have high potential. Nevertheless, this new type of THz source requires further optimization for its use in various applications. The following section discusses a promising technique that can significantly improve THz pulsed emission.

4.3 Double-Pump Technique: Let THz Pulse be Even Stronger

The method of single-color double-pulse excitation is known to be promising for increasing the efficiency of optical-to-THz conversion in the case of filamentation in air [39–41]. The second pulse does not need to waste energy on plasma formation, as it immediately enters the pre-ionized medium, which is the microplasma induced by the first pulse.

To implement double-pulse excitation in the case of a liquid target, Michelson and Mach-Zehnder interferometers are used as a system for forming two successive pulses. The characteristics of the optical pump radiation in each arm are easier to control with the Mach-Zehnder interferometer.

A typical experimental setup with a Mach-Zehnder interferometer is presented in Fig. 4.12. Dispersive media can be used in the arms of the interferomcter to change the duration of the reference and signal pulses. For example, in [42] the duration is changed using quartz plates 2–10 cm thick. It should be noted that the use of dispersion media of the same material in two arms avoids discrepancies in the energy characteristics of

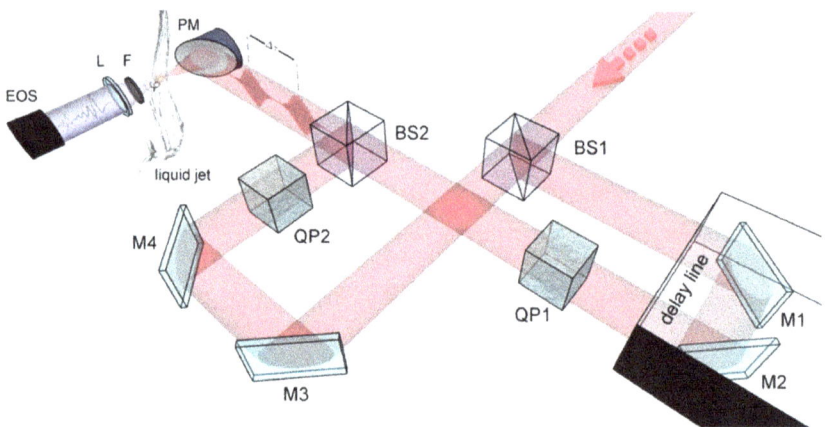

▲ Fig. 4.12.   Experimental scheme of double excitation of plane-parallel liquid jets based on Mach-Zehnder interferometer. Reprinted from Ref. [42] with permission.

the reference and signal pulses. The pulse duration can be controlled by a simple second-order autocorrelator.

Fig. 4.13 demonstrates the experimental results of the enhancement of THz radiation in the case of single-color laser filamentation using double-pulse excitation of a plane-parallel water jet for pulses of the same duration and energy. The temporal forms of THz signals generated in a water jet in

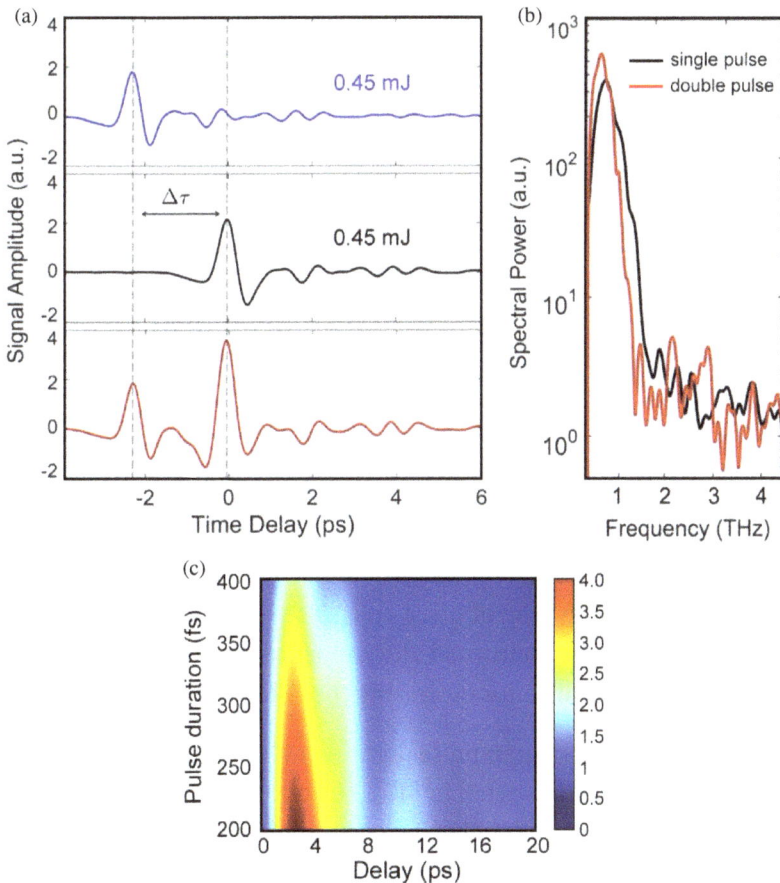

▲ Fig. 4.13.   (a) The temporal forms of THz signals generated during single-color double-pulse excitation of a water jet with a temporal delay $\Delta\tau = 2.3$ ps. (b) The corresponding spectra. (c) Energy dependence of THz waves on the temporal delay $\Delta\tau$ between two pump pulses during propagation in water jet with thickness of 100 μm. Reprinted from Ref. [45] with permission.

the case of double-pulse excitation are shown in Fig. 4.13(a). The figure clearly demonstrates the enhancement of the signal pulse with respect to the reference one. The corresponding THz spectra (Fig. 4.13(b)) are obtained using the Fourier transform.

The general form of the dependences of the enhancement of THz pulse energy on the time delay between two collinear pulses does not differ with a change in the pump duration (Fig. 4.13(c)). It turns out that the maximum enhancement value is obtained when the signal pulse reaches the interaction point approximately 2–4 ps after the reference. This delay corresponds to the time required to create the pre-ionization condition, which involves sufficient electron density, no plasma reflection of the pump pulse [43], and relaxation considering the relaxation time of plasma formation estimated in [42, 44].

Fig. 4.14 depicts the result of numerical simulation of the energy enhancement of THz waves during double-pulse excitation of the liquid. After the interaction of two pulses with a duration of 200 fs, an energy of 450 μJ, and a time delay of 6 ps with the medium, two THz signals are generated at the output (Fig. 4.14(a)). The delay curve is also provided for three different times of electron population relaxation τ_p and is shown in Fig. 4.14(b)–(c).

The comparison of the numerical simulation with the experiment revealed the relaxation time of quasi-free electrons characteristic of water to be about 3 ps, and confirmed the influence of the residual electron density on the mechanism for increasing the optical-to-THz conversion.

The influence of the pump pulse duration on the efficiency in a double-pulse scheme is studied in detail in [42]. Fig. 4.15 shows the comparison of optimal values of the pump pulse duration in the case of single-pulse and double-pulse excitation. In the case of single-pulse excitation (Fig. 4.15(a)), the maximum THz energy is obtained for 200 fs pump pulse duration. For double-pulse excitation with a 2 ps time delay, the optimal value shifts to 100–150 fs (Fig. 4.15(b)). Moreover, in this case, a 20-fold increase is achieved, which is a noticeable improvement.

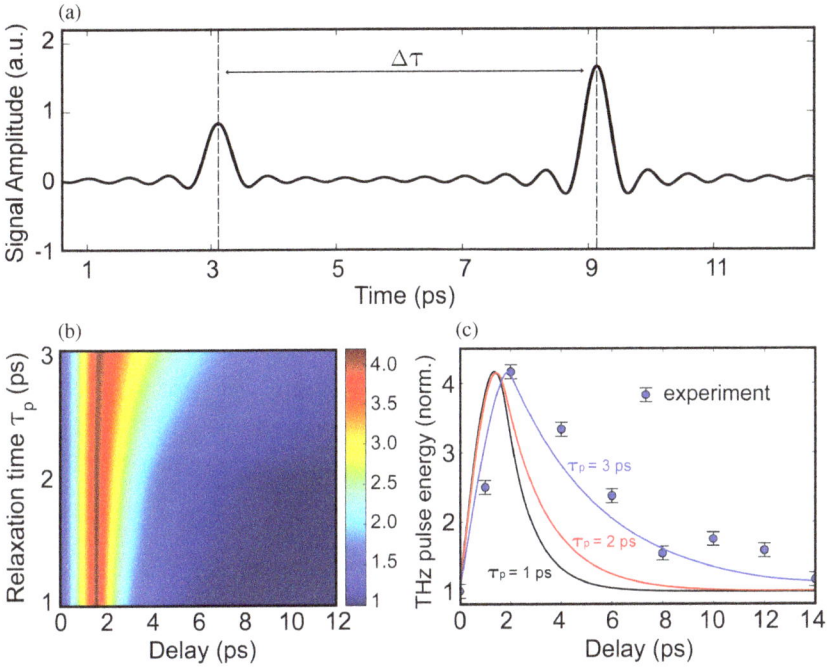

▲ Fig. 4.14.   (a) Numerically simulated THz waveforms generated while launching two collinear 200 fs pulses with energy of 450 μJ and 6 ps temporal delay. (b) The dependence of the energy enhancement of THz waves in the simulation of a single-color double-pulse water jet excitation with the population relaxation time values from 1 to 3 ps. (c) The correspondence of the experimental data (dots) to the result of numerical simulation for $\tau_p = 3$ ps. Reprinted from Ref. [45] with permission.

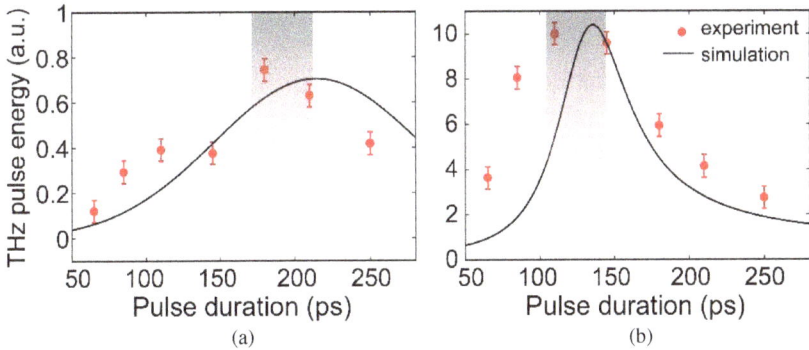

▲ Fig. 4.15.   THz radiation energy dependence on the pump pulse duration. The case of (a) single-pulse and (b) double-pulse excitation of a flat liquid jet are compared. Red dots indicate the experimental results, black solid lines are for the numerical ones. Reprinted from Ref. [42] with permission.

Thus, the first observation is the advantage of shorter pulses in experiments with double-pulse excitation. The presence of the optimal value in the duration in experiments with single-pulse pumping was previously interpreted as a combined effect of the exponential growth of the electron density due to cascade ionization and damping of the pulse energy with increasing pulse duration. The case of double-pulse pumping is different. It can be assumed that the optimal duration experiences a shift towards lower duration values, since the important point here is not the effective ionization of the molecules, but rather the intense interaction with the pre-ionized medium.

Fig. 4.16(a) shows the results of measuring the energy of the generated THz radiation with a change in the ratio of the durations of the reference and signal pulses.

It is readily observed that the characteristic enhancement region around the ratio of the duration of the reference pulse to the signal pulse is 150:150 fs. At the same time, when longer reference pulses generate the initial electron density and are followed by shorter signal pulses, this results in achieving the maximum energy of the THz wave; the enhancement region is located in the lower part of the contour. The numerical simulation (Fig. 4.16(b)) reveals similar trends.

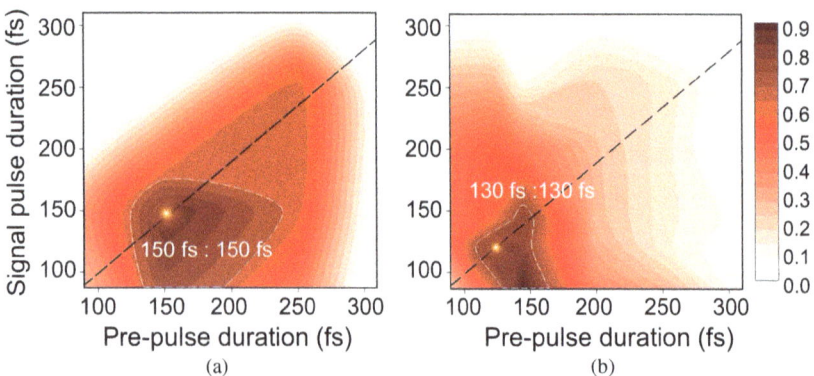

▲ Fig. 4.16. The dependence of the THz radiation energy on the reference pulse/signal pulse duration ratio. (a) Experimentally obtained and extrapolated results. (b) The result of numerical simulations. The dashed line represents the region of the most efficient generation of THz waves. Reprinted from Ref. [42] with permission.

The enhancement of THz waves energy during double-pulse excitation of a liquid jet can be interpreted as follows. The efficiency of generating THz radiation depends on the photocurrent and, accordingly, depends on the induced electron density. Initially, microplasma is formed and THz waves are generated due to weak photocurrents and the nonlinearity of bound electrons. Then, the interaction of the pump pulse with the plasma increases the THz field strength. Because photocurrent is more critical in this case, shorter-duration pulses with high peak power are preferred. This agrees with the expression for the evolution of the current density in Eq. (4.1). The current is proportional to E^3 and the number of quasi-free electrons, that is, the dependence on the peak intensity, is stronger in this case.

A similar trend was also revealed when measuring the dependence of the THz radiation energy on the ratio of the reference pulse energy to the signal one (Fig. 4.17).

The comparison of the experimental points and numerical simulation results in the conclusion that there is an advantage of reference pulses with lower energy and signal pulses with higher energy.

Ponomareva *et al.* [42] demonstrates the results of the THz pulse enhancement in various liquid media, excited by reference and signal pulses

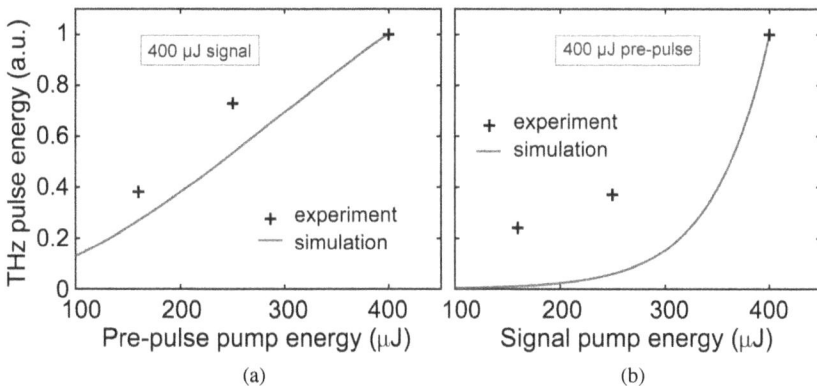

▲ Fig. 4.17. (a) Dependence of the energy of pulsed THz radiation on the energy of the reference pump pulse at a fixed value of the energy of the signal pulse, and (b) inverse dependence. Experimental data are represented by crosses, while numerically simulated data are shown by the green lines. Reprinted from Ref. [46] with permission.

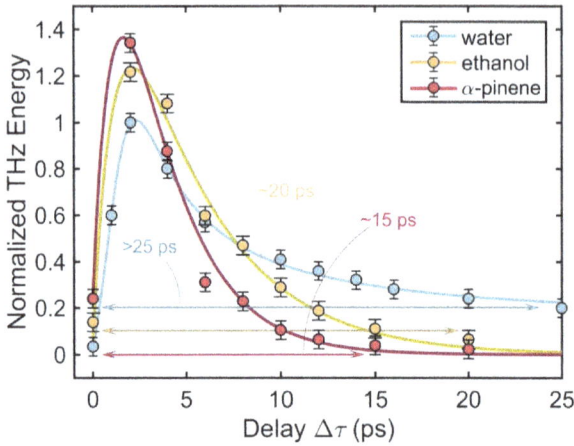

▲ Fig. 4.18. THz wave enhancement dependences on the temporal delay between two collinear 150 fs pulses. The measurements are implemented for water (blue), ethanol (yellow), and α-pinene (red). Reprinted from Ref. [42] with permission.

with a duration of 150 fs and an energy of 450 µJ. The results are normalized to the maximum value for water measured under the same experimental conditions.

Interestingly, the gain curve width in Fig. 4.18 is a characteristic fingerprint for each liquid. In [44], the authors proved experimentally that this width can be interpreted as the plasma lifetime. In addition, a comparison of the obtained curves allows us to conclude that it is advantageous to use α-pinene as a liquid target for generating THz radiation. The maximum value for the efficiency of optical-to-THz conversion, in this case, is about 0.1%.

4.4 Conclusion

Various works are devoted to experimental studies on the generation of THz radiation during laser filamentation in different liquids, including the efficiency of optical-to-THz conversion; the results are given in Tables 4.2–4.4.

▼ Table 4.2. Experimental results of the efficiency of optical-to-THz conversion, spectral range and pulse duration of THz radiation during filamentation in liquids.

Liquid	Polar?	Optical-to-THz Conversion Efficiency		Spectral Width, THz	Pulse Duration, ps (by 1/e level)
		Single Pump	Double Pump		
α-pinene	No	6×10^{-5}	$\sim 10^{-3}$	0.05–1.20	1.9–2.0
Acetone	No	$\sim 10^{-4}$	—	0.05–1.30	1.7–1.8
Distilled water	Yes	1.5×10^{-5}	3×10^{-4}	0.05–1.20	1.9–2.0
Isopropyl	Yes	4×10^{-5}	$\sim 2 \times 10^{-4}$	0.05–1.00	2.2–2.4
Heavy water	Yes	$\sim 10^{-5}$	0.5×10^{-4}	0.05–1.00	2.2–2.4
Ethanol	Yes	2.5×10^{-5}	5×10^{-4}	0.05–1.25	1.8–1.9
Ethylene glycol	Yes	3×10^{-5}	6×10^{-4}	0.05–1.15	2.0–2.2
Hydrogen peroxide	Yes	1×10^{-5}	2×10^{-4}	0.05–1.10	2.1–2.3
p-xylene	No	3.5×10^{-5}	—	0.05–1.40	1.4–1.5

▼ Table 4.3. Experimental results of the efficiency of optical-to-THz conversion, spectral range and pulse duration of THz radiation during filamentation in liquid solutions.

Liquid Solution	Content of the Solution in the Liquid	Optical-to-THz Conversion Efficiency		Spectral Width, THz	Pulse Duration, ps (by 1/e level)
		Single Pump	Double Pump		
Sodium carbonate solution	4 kH	10^{-5}	—	0.05–1.00	2.2–2.4
	10 kH	7×10^{-6}			
	14 kH	4×10^{-6}			
	33 kH	2×10^{-6}			
Acid solutions	3 pH	8×10^{-6}	—	0.05–1.00	2.2–2.4
	4 pH				
	6.5 pH				
	7 pH				
Saline solution	0%	1.5×10^{-5}	—	0.05–1.20	1.9–2.0
	3%			0.05–1.10	2.1–2.3
	7%				
	15%				
	20%				

(Continued)

▼ Table 4.3.　(*Continued*)

Liquid Solution	Content of the Solution in the Liquid	Optical-to-THz Conversion Efficiency		Spectral Width, THz	Pulse Duration, ps (by 1/e level)
		Single Pump	Double Pump		
Spirit solution	0%	1.5×10^{-5}	—	0.05–1.20	1.9–2.0
	25%	1.7×10^{-5}			
	50%	1.9×10^{-5}			
	75%	2.1×10^{-5}			
	100%	2.5×10^{-5}		0.05–1.25	1.8–1.9

▼ Table 4.4.　Experimental results of the efficiency of optical-to-THz conversion, spectral range and pulse duration of THz radiation during filamentation in food liquid solutions.

Liquid Solution	Difference	Optical-to-THz Conversion Efficiency		Spectral Width, THz	Pulse Duration, ps (by 1/e level)
		Single Pump	Double Pump		
Coca-Cola	With sugar	5×10^{-6}	—	0.05–0.85	2.7–2.9
	Sugar free				
Wine	Red	7×10^{-6}	—	0.05–1.00	2.2–2.4
	White				
Vodka	1	1.7×10^{-5}			
	2	2.0×10^{-5}			
	3	1.9×10^{-5}	—	0.05–1.25	1.8–1.9
	4	1.85×10^{-5}			
Milk	0.25%				
	0.5%				
	1.5%	9×10^{-6}	—	0.05–1.00	2.2–2.4
	2.5%				
	2.8%				
	6%				

As can be seen from the tables, the technique of double-pulse excitation of the medium allows increasing the energy of the generated THz pulse by an order of magnitude on average. Adding various liquids to water does not lead to a significant change in the efficiency of conversion of optical pumping

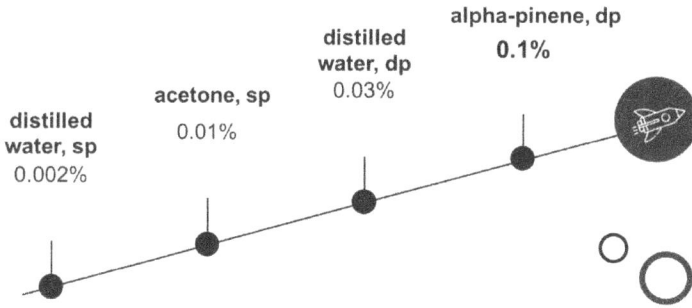

▲ Fig. 4.19. The dependence of the efficiency of optical-THz conversion during laser filamentation in polar and nonpolar liquids. sp is single-pulse excitation, and dp is double-pulse excitation.

into THz radiation, which is due to the absence of a significant effect from the studied impurities on the processes of ionization of the substance and the generation of photocurrent. Currently, the most effective liquid for generating THz radiation is α-pinene using the double pumping method. The general dependence of the change in the efficiency of optical-to-THz conversion when using various methods and liquids is shown in Fig. 4.19.

References

1. Kennedy P. K. (1995). A first–order model for computation of laser–induced breakdown thresholds in ocular and aqueous media. I. Theory, IEEE Journal of Quantum Electronics, 31(12), pp. 2241–2249.
2. Hammer D. X., Thomas R. J., Noojin G. D., Rockwell B. A., Kennedy P. K. & Roach W. P. (1996). Experimental investigation of ultrashort pulse laser–induced breakdown thresholds in aqueous media, IEEE Journal of Quantum Electronics, 32(4), pp. 670–678.
3. Fullagar W., Harbst M., Canton S., Uhlig J., Walczak M., Wahlström C. G. & Sundström V. (2007). A broadband laser plasma x–ray source for application in ultrafast chemical structure dynamics, Review of Scientific Instruments, 78(11), pp. 115105.
4. Stokholm J., Blaser M. J., Thorsen J., Rasmussen M. A., Waage J., Vinding R. K., ... & Bisgaard H. (2018). Maturation of the gut microbiome and risk of asthma in childhood, Nature Communications, 9(1), pp. 1–10.

5. Vasa P., Dharmadhikari J. A., Dharmadhikari A. K., Sharma R., Singh M. & Mathur D. (2014). Supercontinuum generation in water by intense, femtosecond laser pulses under anomalous chromatic dispersion, Physical Review A, 89(4), pp. 043834.

6. Jin Q., E Y., Williams K., Dai J. & Zhang X.-C. (2017). Observation of broadband terahertz wave generation from liquid water, Applied Physics Letters, 111(7), pp. 071103.

7. Tcypkin A. N., Ponomareva E. A., Putilin S. E., Smirnov S. V., Shtumpf S. A., Melnik M. V., ... & Zhang X.-C. (2019). Flat liquid jet as a highly efficient source of terahertz radiation, Optics Express, 27(11), pp. 15485–15494.

8. Watanabe A., Saito H., Ishida Y., Nakamoto M. & Yajima T. (1989). A new nozzle producing ultrathin liquid sheets for femtosecond pulse dye lasers, Optics Communications, 71(5), pp. 301–304.

9. Ismagilov A. O., Ponomareva E. A., Zhukova M. O., Putilin S. E., Nasedkin B. A. & Tcypkin A. N. (2021). Liquid jet–based broadband terahertz radiation source, Optical Engineering, 60(8), pp. 082009.

10. Ponomareva E. A., Putilin S. E., Tcypkin A. N., E Y., Kozlov S. A. & Zhang, X.-C. (2019). Comparison of various liquids as sources of terahertz radiation from one–color laser filament, Proceedings of Infrared, Millimeter-Wave, and Terahertz Technologies VI, SPIE, 11196, pp. 23–29.

11. Naito H., Ogawa Y., Suzuki T. & Kondo N. (2011). Milk inspection by THz attenuated total reflectance (Thz–ATR) spectroscopy, Proceedings of the American Society of Agricultural and Biological Engineers, pp. 1.

12. Wilkes Z. W., Varma S., Chen Y. H., Milchberg H. M., Jones T. G. & Ting A. (2009). Direct measurements of the nonlinear index of refraction of water at 815 and 407 nm using single–shot supercontinuum spectral interferometry, Applied Physics Letters, 94(21), pp. 211102.

13. Smith W. L., Liu P. & Bloembergen N. (1977). Superbroadening in H 2 O and D 2 O by self–focused picosecond pulses from a YAlG: Nd laser, Physical Review A, 15(6), pp. 2396.

14. Liu J., Schröder H., Chin S. L., Li R. & Xu Z. (2005). Nonlinear propagation of fs laser pulses in liquids and evolution of supercontinuum generation, Optics Express, 13(25), pp. 10248–10259.

15. Markowicz P. P., Samoc M., Cerne J., Prasad P. N., Pucci A. & Ruggeri G. (2004). Modified Z–scan techniques for investigations of nonlinear chiroptical effects, Optics Express, 12(21), pp. 5209–5214.

16. Huang S. S. S. & Freeman G. R. (1977). Effect of density on the total ionization yields in x–irradiated argon, krypton, and xenon, Canadian Journal of Chemistry, 55(11), pp. 1838–1846.
17. Faubel M., Steiner B. & Toennies J. P. (1997). Photoelectron spectroscopy of liquid water, some alcohols, and pure nonane in free micro jets, The Journal of Chemical Physics, 106(22), pp. 9013–9031.
18. Truong S. Y., Yencha A. J., Juarez A. M., Cavanagh S. J., Bolognesi P. & King G. C. (2009). Threshold photoelectron spectroscopy of H2O and D2O over the photon energy range 12–40 eV, Chemical Physics, 355(2–3), pp. 183–193.
19. Holmes J. L. & Lossing F. P. (1982). Towards a general scheme for estimating the heats of formation of organic ions in the gas phase. Part II. The effect of substitution at charge–bearing sites, Canadian Journal of Chemistry, 60(18), pp. 2365–2371.
20. Bowen R. D. & Maccoll A. (1984). Low energy, low temperature mass spectra of some small saturated alcohols and ethers, Organic Mass Spectrometry, 19(8), pp. 379–384.
21. Al–Joboury M. I. & Turner D. W. (1964). 851. Molecular photoelectron spectroscopy. Part II. A summary of ionization potentials, Journal of the Chemical Society (Resumed), pp. 4434–4441.
22. Weber M. J. (2018). *Handbook of Optical Materials.* CRC Press.
23. Egorov G. I., Makarov D. M. & Kolker, A. M. (2010). Densities and volumetric properties of ethylene glycol+ dimethylsulfoxide mixtures at temperatures of (278.15 to 323.15) K and pressures of (0.1 to 100) MPa, Journal of Chemical and Engineering Data, 55(9), pp. 3481–3488.
24. Chu K. Y. & Thompson A. R. (1962). Densities and refractive indices of alcohol–water solutions of n–propyl, isopropyl, and methyl alcohols, Journal of Chemical and Engineering Data, 7(3), pp. 358–360.
25. Tavares Sousa A. & Nieto de Castro C. A. (1992). Density of α–pinene, β–pinene, limonene, and essence of turpentine, International Journal of Thermophysics, 13(2), pp. 295–301.
26. Tan N. Y., Li R., Bräuer P., D'Agostino C., Gladden L. F. & Zeitler J. A. (2015). Probing hydrogen–bonding in binary liquid mixtures with terahertz time–domain spectroscopy: a comparison of Debye and absorption analysis, Physical Chemistry Chemical Physics, 17(8), pp. 5999–6008.

27. Chong J., Maeng I., Shin H. J. & Son J. H. (2009). Terahertz characteristics of liquid D2O in H2O. AIP Conference Proceedings, 1119(1), pp. 209–209.

28. George D. K. & Markelz A. G. (2012). Terahertz Spectroscopy of Liquids and Biomolecules, in *Terahertz Spectroscopy and Imaging*, Springer, pp. 229–250.

29. Huang S., Ashworth P. C., Kan K. W., Chen Y., Wallace V. P., Zhang Y. T. & Pickwell–MacPherson E. (2009). Improved sample characterization in terahertz reflection imaging and spectroscopy, Optics Express, 17(5), pp. 3848–3854.

30. Dey I., Jana K., Fedorov V. Y., Koulouklidis A. D., Mondal A., Shaikh M., ... & Kumar, G. R. (2017). Highly efficient broadband terahertz generation from ultrashort laser filamentation in liquids, Nature Communications, 8(1), pp. 1–7.

31. Bugay A. N. & Sazonov S. V. (2010). The generation of terahertz radiation via optical rectification in the self–induced transparency regime, Physics Letters A, 374(8), pp. 1093–1096.

32. Vysotina N. V., Rozanov N. N. & Semenov V. E. E. (2006). Extremely short pulses of amplified self–induced transparency, Journal of Experimental and Theoretical Physics Letters, 83(7), pp. 279–282.

33. Balakin A. V., Coutaz J. L., Makarov V. A., Kotelnikov I. A., Peng Y., Solyankin P. M., ... & Shkurinov A. P. (2019). Terahertz wave generation from liquid nitrogen, Photonics Research, 7(6), pp. 678–686.

34. Stumpf S. A., Korolev A. A. & Kozlov S. A. (2007). Few–cycle strong light field dynamics in dielectric media, Proceedings of Laser Optics 2006: Superintense Light Fields and Ultrafast Processes, SPIE, 6614, pp. 59–70.

35. Kozlov S. A. & Samartsev V. V. (2013). *Fundamentals of Femtosecond Optics*. Elsevier.

36. Mazurenko Y. T., Spiro A. G., Putilin S. E., Beliaev A. G. & Verkhovskij E. B. (1995). Time–to–space conversion of fast signals by the method of spectral nonlinear optics, Optics Communications, 118(5–6), pp. 594–600.

37. Ponomareva E. A., Stumpf S. A., Tcypkin A. N. & Kozlov S. A. (2019). Impact of laser–ionized liquid nonlinear characteristics on the efficiency of terahertz wave generation, Optics Letters, 44(22), pp. 5485–5488.

38. Lindner J., Unterreiner A. N. & Vöhringer,P. (2006). Femtosecond relaxation dynamics of solvated electrons in liquid ammonia, ChemPhysChem, 7(2), pp. 363–369.
39. Xie X., Xu J., Dai J. & Zhang, X.-C. (2007). Enhancement of terahertz wave generation from laser induced plasma, Applied Physics Letters, 90(14), pp. 141104.
40. Kim K. Y., Glownia J. H., Taylor A. J. & Rodriguez G. (2007). Terahertz emission from ultrafast ionizing air in symmetry–broken laser fields, Optics Express, 15(8), pp. 4577–4584.
41. Mori K., Hashida M., Nagashima T., Li D., Teramoto K., Nakamiya Y., ... & Sakabe S. (2017). Directional linearly polarized terahertz emission from argon clusters irradiated by noncollinear double–pulse beams, Applied Physics Letters, 111(24), pp. 241107.
42. Ponomareva E. A., Ismagilov A. O., Putilin S. E., Tsypkin A. N., Kozlov S. A. & Zhang, X.-C. (2021). Varying pre–plasma properties to boost terahertz wave generation in liquids, Communications Physics, 4(1), pp. 1–7.
43. Sarpe–Tudoran C., Assion A., Wollenhaupt M., Winter M. & Baumert T. (2006). Plasma dynamics of water breakdown at a water surface induced by femtosecond laser pulses, Applied Physics Letters, 88(26), pp. 261109.
44. Ponomareva E., Ismagilov A., Putilin S. & Tcypkin A. N. (2021). Plasma reflectivity behavior under strong subpicosecond excitation of liquids, APL Photonics, 6(12), pp. 126101.
45. Ponomareva E., Tcypkin A., Smirnov S., Putilin S., E Y., Kozlov S. & Zhang, X.-C. (2019). Double-pump technique–one step closer towards efficient liquid-based THz sources, Optics Express, 27(22), pp. 32855–32862.
46. Ponomareva E., Ismagilov A., Putilin S., Tsypkin A. & Kozlov S. (2020). Laser and matter properties effect on the enhancement of THz waves energy generated during liquid jets double-pulse excitation, Fourth International Conference on Terahertz and Microwave Radiation: Generation, Detection, and Applications, SPIE, 11582, pp. 48–52.

https://doi.org/10.1142/9789811265648_0005

Chapter 5

Giant Nonlinearity of Liquids in the
Terahertz Frequency Range

Anton Tcypkin, Melnik Maksim, Zhukova Maria, Kozlov Sergei

ITMO University, Russia

5.1 Introduction

Radiophysics and optics are currently contributing to the development of the terahertz (THz) frequency range of electromagnetic radiation. Such active involvement is justified by the discovery of wide applications of THz radiation in biology, medicine, security systems, and wireless information transmission devices [1–5]. Until recently, the systems and devices developed in THz photonics were mainly based on phenomena of linear THz optics [6–8]. However, the last decade has marked the appearance of sources of pulsed THz radiation with energy in a single pulse up to 125 µJ and peak intensities up to 10^{13} W/cm² [9]. Nowadays, sources of high-power THz radiation based on the optical rectification of femtosecond pulses in $LiNbO_3$ crystals are widely used in many scientific laboratories [10–12]. For example, such sources were used by the group of Prof. Nelson to observe nonlinear effects [13]. It is THz sources that were used to obtain the experimental results discussed in this chapter.

Progress in the development of high-power pulsed THz radiation sources allowed researchers to start extensive research in the field of nonlinear THz optics [14–16]. To date, several reviews in this area have already appeared [17–20]. Unexpectedly, the analysis of basic problems of nonlinear optics in the THz frequency range surprised the scientific community. For instance, the nature of the refractive index nonlinearity of materials needs to be specified in this spectral range. New features of the self-action of THz waves in nonlinear media have been revealed. Therefore, new approaches are required to analyze promising applications for nonlinear optics phenomena of THz radiation [21].

One of the most fascinating surprises is the discovery of a giant low-inertia nonlinearity of the refractive index of some materials in the THz spectral range. Dolgaleva et $al.$ [22] predicted that the coefficient of the low-inertia nonlinear refractive index n_2 for certain crystals in the THz spectral range can be several orders higher than the values of this coefficient of the same materials in the visible and near-infrared (IR) spectral ranges. At the same time, it was shown that such a large nonlinearity of media has an oscillatory nature, and the dominant low-inertia mechanism for the nonlinearity of refractive index in the field of THz waves is the anharmonic vibrations of atoms in molecules of a substance. Self-phase modulation was first experimentally observed in THz range in [23]. The observations demonstrated significant radiation-induced changes in the refractive index of the material under study. It turned out that n_2 of a lithium niobate crystal in the field of pulsed THz radiation is four orders higher than in the visible and near-IR ranges [24, 25]. However, the largest values of n_2, which are millions of times higher than in the visible and near-IR spectral ranges, were experimentally found in liquids [26–30]. For water, the nonlinear refractive index coefficient is $n_2 = 7 \times 10^{-10}$ cm²/W. Notably, it was experimentally shown that the time for establishing a nonlinear response in a liquid is 1 ps or less, i.e., the mechanism of its gigantic nonlinearity is low inertia. Such a high and low-inertia nonlinearity of some materials in the THz spectral range opens promising prospects for the development of ultrafast THz photonics based on the wave self-action effects.

This chapter discusses the features of methods for n_2 estimation of materials in the field of pulsed THz radiation, presents the results of n_2

estimation for various liquids, and theoretically analyzes the nature of the giant low-inertia nonlinearity of liquids in the THz spectral range.

5.2 Features of Techniques to Estimate Nonlinear Refractive Index Coefficient in the THz Range

To begin with, let us discuss methods that allow parameter evaluation of materials that characterize the nonlinearity of their polarization response under the influence of THz radiation. In general, the observation of any nonlinear effect makes it possible to estimate such parameters. Thus, the analysis of the change in the temporal shape of the THz field due to its self-phase modulation in a lithium niobate crystal allowed the authors [24] to determine the coefficient of the nonlinear refractive index of the crystal; they found its value equal to $n_2 = 4 \times 10^{-11}$ cm²/W.

Recall that this coefficient is one of the most important characteristics of the nonlinear properties of optical materials. It determines the change in the refractive index of the material in the field of high-intensity radiation

$$n(\omega) = n_0(\omega) + n_2 I, \qquad (5.1)$$

where n_0 is the linear refractive index of the medium, I is the radiation intensity, and n_2 is nonlinear refractive index coefficient (usually in SI [31]).

The essence of the measurement method used in [24] asserts that the time for establishing the nonlinearity of the refractive index of the medium is shorter than the pulse duration, which is about 1 ps. As noted in the introduction, this is important, since the low-inertia nonlinearity of the refractive index is promising for the development of nonlinear photonic systems that can compete with electronics in terms of response time.

Experimental and theoretical research on the disappearance of radiation generation at tripled frequencies and appearance of radiation generation at quadruple frequencies in the field of a single-cycle THz wave allowed the authors [32] to determine n_2 of lithium niobate crystals as $n_2 = (8 + 1) \times 10^{-11}$ cm²/W. The estimations of n_2 for the same crystal by qualitatively different methods appeared to have close values. Thus, the fact of a very large and low-inertia nonlinearity of the lithium niobate crystal is likely to be reliable. Measurement of n_2 of liquids in the THz frequency range can be performed using various techniques, for example, based on full phase analysis [30], pump-probe spectroscopy [33] or nonlinear transmission on certain spectral lines [34].

One of the most widely used techniques for n_2 estimation of optical materials in the visible and near-IR ranges of the spectrum is z-scan [35]. In this chapter the z-scan method was used to estimate the n_2 values of liquids in the THz range of the spectrum given. Let us consider the features of the method for pulsed broadband radiation in the THz spectral range.

The z-scan technique is based on changing the divergence of a light beam focused by a mirror or lens after it passes through the focus, in the vicinity of which a layer of the studied nonlinear medium is located. This layer moves along the beam propagation axis. A layer of a nonlinear medium acts as an additional thin lens with the intrinsic focal length depending on how far it is shifted from the main focus. The layer displacement along the beam axis near the focus leads to a change in distribution of the beam field in the far field region. The fraction of radiation energy, which enters a fixed aperture in the far field region and is recorded by a photodetector, depends on the displacement of the nonlinear medium from the focus. The resulting characteristic curve of the z-scan method has a peak and valley of the diaphragm transmission; the peak-to-valley ratio can be used to determine the coefficient of the nonlinear refractive index.

This technique is strictly justified for monochromatic radiation [35] but can also be used for femtosecond radiation with a fairly wide spectrum [36]. However, the possibility of using the z-scan method in the THz frequency range requires additional justification, since the spectrum of pulsed THz sources is even wider than that of the femtosecond [37], and its electric field can have only one oscillation [38].

To substantiate the possibility of using the z-scan method for broadband few-cycle THz pulses, a paper [39] compares the results of numerical

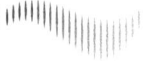

simulation of the z-scan method for ZnSe crystal using broadband THz radiation with an analytical model of the method for monochromatic radiation. It should be noted that the pulse duration does not appear in the analytical model and, therefore, the analytical curve is independent of the given pulse characteristic.

Let us consider the propagation of a spherical THz beam over a distance corresponding to two focal lengths. The resulting THz pulses are detected through an aperture for different sample positions on the optical propagation axis. A graphical interpretation of the numerical simulation is shown in Fig. 5.1(a). In the figure, the minimum of the electric field is blue, and the maximum is red. For each position of the ZnSe crystal, the THz pulse propagates in the air, the crystal itself, and then travels the remaining distance in the air before reaching the detector after passing through the aperture.

Fig. 5.1(b)–(i) present the examples of single- and multi-cycle pulses for a center wavelength of 300 μm. To show details on the transverse pulse size scale, wavelength fractions calculated as the ratio r/λ_0 are used as units. A 1 ps pulse actually consists of 1.5 oscillations, while a true single-cycle pulse corresponds to a duration of 0.3 ps [40].

However, Fig. 5.1(e) shows that the spectrum of a true single-cycle pulse shifts to higher values. For numerical simulations, this means that the maximum wavelength of radiation is reduced and, consequently, the waist radius also decreases. In turn, this increases both the peak intensity in the caustic by a factor of 1.5 and peak-to-valley ratio of the z-scan curve. THz pulses generated experimentally are close to true single cycles [41]. Therefore, it is crucial to consider the above nuances in the course of numerical simulation to match with experimental results as closely as possible. It should be noted that although the focal length is much larger than the transverse size of the THz beam (which corresponds to the case of paraxial propagation), the curvature of the wavefront of true single-cycle pulses is very strong. The latest analysis of nonlinear effects for non-paraxial beams of single-cycle THz pulses [42] shows that the paraxial approximation well describes the dynamics of the transverse field

(a)

P   S   A
                                                                                                              D

z−   0   z+

(b)

r/λ_0 (a.u.)

(c)

r/λ_0 (a.u.)

(d)

E (a.u.)

(e)

$|G|$ (a.u.)

ω_{norm} (a.u.)

(f)

r/λ_0 (a.u.)

(g)

r/λ_0 (a.u.)

(h)

E (a.u.)

(i)

$|G|$ (a.u.)

ω_{norm} (a.u.)

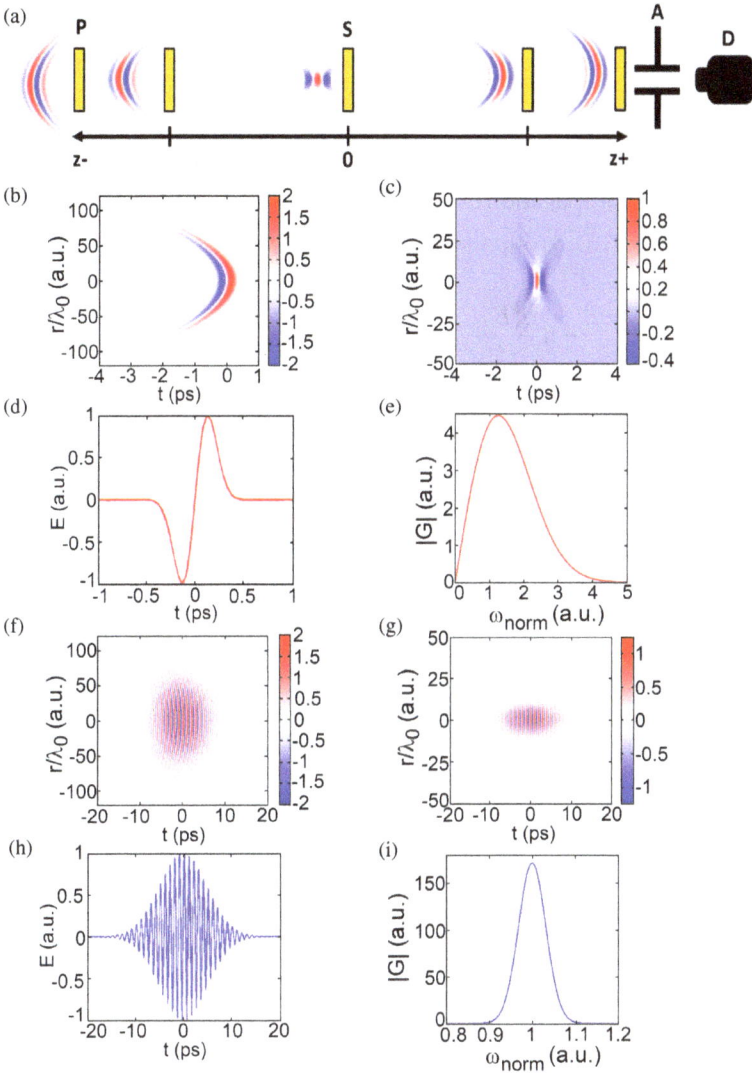

▲ Fig. 5.1.   (a) Graphical representation of the z-scan method for pulsed THz radiation propagating from z⁻ to z⁺: for each position of the crystal the THz pulse propagates in the air, in the crystal itself, and then travels the remaining distance in the air before being finally detected at point z⁺. The propagation distance corresponds to two focal lengths. Here P is the initial position of the crystal, S is the focal point, A is the position of the aperture and D is the detector. (b, c) Single- and (f, g) multi-cycle THz pulses. (d, h) The cross-sections of their electric field profiles and (e, i) spectra, respectively. Reprinted from Ref. [86] with permission.

component and strongly focused non-paraxial THz beams. The appearance and nontrivial dynamics of the longitudinal component of the field are different in this case.

The described numerical simulation of the z-scan method is based on the equations for the propagation of an intense light pulse in a waveguide dielectric medium with normal group dispersion and nonresonant nonlinearity [43]:

$$\frac{\partial E}{\partial z} - a\frac{\partial^3 E}{\partial \tau^3} + gE^2\frac{\partial E}{\partial \tau} = \frac{c}{2N_0}\Delta_\perp \int_{-\infty}^{\tau} E d\tau',$$

(5.2)

where the second term on the left describes the dispersion of the linear polarization response, the third describes the nonlinear medium response, and the term on the right describes the diffraction of an extremely short pulse. In this formula, z is the direction of propagation, $\Delta_\perp$ is the transverse Laplacian, a is the coefficient from the linear refractive index dispersion expression: $n_0(\omega) = N_0 + ac\omega^2$, where N_0 is the refractive index at the "zero" frequency, E is the electric field of THz wave, and $E(z=0,r,t) = E_0 \times exp(-r^2/r_0^2) \times exp(-t^2/\tau_0^2) \times sin(\omega_0 t) \times R(x,y)$ is the electric field for a spherical THz wave at the input of the medium. Here, $r = \sqrt{x^2 + y^2}$ is the beam coordinate in the aperture, $r_0 = d/2$ is the input beam radius, τ_0 is the pulse duration, ω_0 is the central frequency, $R(x,y) = exp(-ik(x^2 + y^2)/2f)$ is the transmission function of a spherical lens with focal length f, $k = n_0\omega/c$ is the wavenumber, and x and y are transverse coordinates. The parameter g characterizes the nonlinear polarization response of electronic nature and determines the instantaneous part of the medium nonlinear refractive index coefficient $n_2 = g \times c/2$ (in CGS units).

The numerical simulation is carried out for the following system parameters: the sample under study - ZnSe (thickness $L = 0.06{-}0.3$ mm), focal length $f = 50$ cm, central wavelength of the THz pulse $\lambda_0 = 0.2{-}0.6$ mm, oscillation period $T_0 = 0.75{-}2$ ps, pulse duration $\Delta t = 0.3{-}10\ T_0$, input beam transverse width $D = 25$ mm, aperture size $A = 1.5$ mm, and peak intensity in the caustic $I_0 = 3.1 \times 10^8$ W/cm². The dispersion parameters are taken from [44] and approximated by the formula $n_0(\omega) = N_0 + ac\omega^2$, where $N_0 = 3$ and

$a = 6 \times 10^{-36}$ s^3/m. The value, $n_2 = 4 \times 10^{-11}$ cm^2/W, taken from [45] is a result of the experimental data evaluation.

In the course of processing data from an array of numbers characterizing the signal at each point of the THz field, the energy is integrated along the time axis for an aperture of radius $r_0 = 0.75$ mm and $t = 2000$ fs using the equation:

$$\int_{r=0}^{r=r_0} \int_{\varphi=0}^{\varphi=2\pi} \int_{\tau=-t}^{\tau=t} E^2 \left(\tau, \varphi, r \right) d\tau \, d\varphi \, dr. \tag{5.3}$$

Fig. 5.2 shows the results of the numerical simulation of typical z-scan curves, which represent the dependence of the transmission through the closed aperture (T) with the sample on the position with coordinate z for the following parameters: central wavelength of THz radiation $\lambda_0 = 0.3$ mm, period of oscillation $T_0 = 1$ ps, peak intensity in the caustic (a) $I_0 = 3.1 \times 10^8$ W/cm^2 and (b) $I_0 = 8.3 \times 10^8$ W/cm^2, pulse duration $\tau_0 = 0.3$, 1, 10 T_0, and ZnSe sample thickness $L = 0.3$ mm. The curves obtained qualitatively correspond to the z-scan curves. For the central wavelength $\lambda_0 = 0.3$ mm, the duration value $\tau_0 = 0.3$ ps corresponds to a single oscillation of the electromagnetic field (see Fig. 5.1(d)). The simulation in Fig. 5.2(b) corresponds to a different intensity value due to the fact described above. The waist diameter of a true single-cycle pulse is two times smaller than for multi-cycle pulses. Focusing such a pulse leads to its transformation into a pulse of 1.5 oscillations, which occurs due to the shift of the maximum frequency (with constant ω_0) of a true single-cycle pulse to the region of higher values (see Fig. 5.1(e)).

To obtain an analytical representation of the z-scan curve for monochromatic radiation (the solid black lines in Fig. 5.2), let us now consider [35]. Transmission through a closed aperture T can be calculated using the following formula:

$$T(z) = \frac{\int_{-\infty}^{+\infty} P_T(\Delta \Phi_0(t)) dt}{S \cdot \int_{-\infty}^{+\infty} P_i(t) dt}, \tag{5.4}$$

(a)

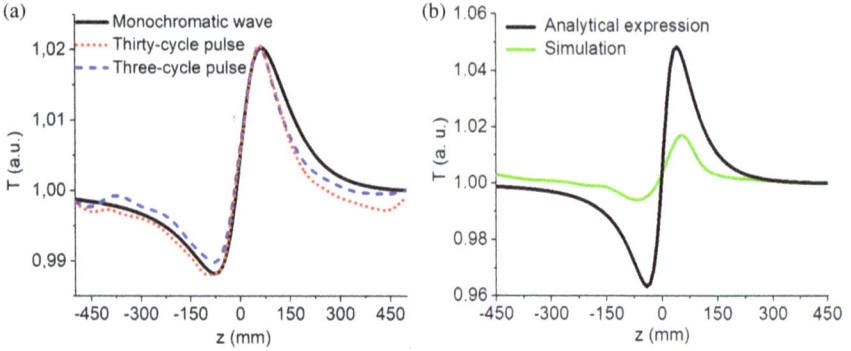

(b)

Fig. 5.2. Dependence of the transmission through a closed aperture T on the position of the crystal z obtained by numerical simulation for pulses with a duration of (a) 1 and 10 T_0 and (b) 0.3 T_0 for a central wavelength $\lambda_0 = 0.3$ mm and a crystal with a thickness of 0.3 mm; the black curve corresponds to the analytical curve for monochromatic THz radiation. Reprinted from Ref. [39] with permission.

where $P_i(t) = \frac{\pi w_0^2 I_0(t)}{2}$ is instantaneous input power; $S = 1 - exp(-2\frac{r_a^2}{w_a^2})$ is linear aperture transmission with w_a denoting the beam radius at the aperture in the linear regime and r_a the aperture radius; $I_0(t)$ is the irradiance within the sample; and w_0 is the beam waist radius. The numerator, in turn, can be calculated as

$$P_T\left(\Delta\Phi_o\left(t\right)\right) = c\varepsilon_o N_o \pi \int_0^{r_a} |E_a(r,t)|^2\, r dr,$$

(5.5)

where c is the speed of light, ε_o is the permittivity of vacuum, N_o is the linear index of refraction, $E_a(r,t) = E(z, r = 0, t)exp(-\frac{\alpha L}{2})\times\sum_{m=0}^{+\infty}\frac{[i\Delta\varphi_0(z,t)]^m}{m!}\frac{w_{m0}}{w_m}exp(-\frac{r^2}{w_m^2} - \frac{ikr^2}{2R_m} + iQ_m)$ is the resultant electric field pattern at the aperture (see [35]), $E(z, r = 0, t) = E_o\, exp\,(-2\frac{t^2}{\tau_0^2})sin(\omega_o t)\frac{w_o}{w(z)}$ is the electric field on the axis, $w(z)$ is the beam radius, $\Delta\varphi_o(z,t) = \frac{\Delta\Phi_0(t)}{1+z^2/z_0^2}$ is the phase shift at the focus, $\Delta\Phi_o(t) = kn_2 I_0(t)L$ is the on-axis phase shift at the focus, and L is the sample length.

The calculation is carried out for the following parameters: absorption coefficient $\alpha = 0.85$ cm^{-1}, sample thickness $L = 0.3$ mm, central radiation frequency $v_0 = 1$ THz ($\lambda_0 = 0.3$ mm), beam radius in the waist $w_0 = 1.4$ mm, aperture radius $r = 1.5$ mm, initial beam radius $w_a = 12.5$ mm, peak radiation intensity in the caustic $I_0 = 3 \times 10^8$ W/cm^2, and nonlinear refractive index coefficient $n_2 = 4 \times 10^{-11}$ cm^2/W.

Note that the shorter the pulse duration, the less the numerically calculated peak-to-valley ratio in the z-scan curve corresponds to the analytical curve and, consequently, the higher the inaccuracy of the obtained results of n_2 estimation. It can be concluded that the z-scan method has the most significant error in determining n_2 with pulses containing less than two oscillations. As mentioned above, for a true single-cycle pulse, there is a change in its duration in the caustic and a decrease in the size of the waist compared to the multi-cycle pulse. In addition, the maximum frequency is shifted to higher values (see Fig. 5.1(e)). All these facts increase the peak intensity in the caustic by a factor of 2.7 comparing to the multi-cycle pulse. Despite this, the peak-to-valley ratio of the true single-cycle pulse is less than that for a two-cycle pulse, as can be seen in Fig. 5.2. The analytical expression gives $n_2 = 4 \times 10^{-11}$ cm^2/W, while the numerical simulation curve provides $n_2 = 1.15 \times 10^{-11}$ cm^2/W. Thus, in this case the inaccuracy of the method is more than 70%.

Fig. 5.3 shows the dependence of the n_2 estimation error from the differential curve obtained by the z-scan numerical simulation on the number of THz pulse periods T_0 in the case of a fixed sample thickness $L = 0.3$ mm for different wavelengths $\lambda_0 = 0.3$ mm and 0.4 mm.

The results demonstrate that the n_2 estimation error increases drastically if the number of pulse cycles decreases, when the pulse has less than two oscillations. The largest error value corresponds to a true single-cycle pulse and is practically independent of the radiation wavelength.

To explain the discrepancy between the simulation results for pulses containing less than two oscillations and analytical calculations for monochromatic THz radiation, we turn to [46]. The authors reveal that the dispersion effects for pulses consisting of only a few field oscillations can be commensurate with high nonlinearity. Thus, dispersion length

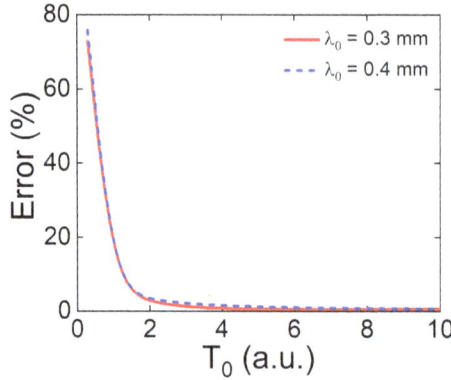

▲ Fig. 5.3.   Dependence of the n_2 estimation error on the number of cycles in a THz pulse for a fixed sample thickness $L = 0.3$ mm and different wavelengths $\lambda_0 = 0.3$ mm and 0.4 mm. Reprinted from Ref. [39] with permission.

L_{disp} is commensurate with nonlinear length L_{nl} or even less than the latter [40].

Knowing the pulse and propagation medium parameters, the dispersion and nonlinear lengths can be calculated. Drozdov *et al.* [40] suggested using the following formulas for few-cycle pulses:

$$L_{disp} = \frac{\pi^2 \lambda_0^2 N_0}{16\Delta n_{disp}}; \; L_{nl} = \frac{\lambda_0^2 N_0}{16\Delta n_{nl}}, \tag{5.6}$$

where $\Delta n_{disp} = ac\omega^2$ is the change in the refractive index at the center wavelength λ_0 due to dispersion, $\Delta n_{nl} = 1/2 \, n_2 I$ is the nonlinearly induced change in the optical refractive index, and $\omega_0 = 2\pi/T_0$ is the center frequency. The dispersion properties of the medium can be expressed through $n_0(\omega) = N_0 + ac\omega^2$, where N_0 is the refractive index at the "zero" frequency, and a is the dispersion coefficient. Thus, for $N_0 = 3$, $a = 6 \times 10^{-36}$ s³/m, $\lambda_0 = 0.3$ mm, $\omega_0 = 6.28 \times 10^{12}$ rad/s, $I_0 = 3.1 \times 10^8$ W/cm², and $n_2 = 4 \times 10^{-11}$ cm²/W. The calculated values of the dispersive and nonlinear lengths are $L_{disp} = 7.81$ mm and $L_{nl} = 9.07$ mm, respectively.

The values obtained exceed the sample thickness, $L = 0.3$ mm. In this regard, it can be assumed that an increase in the sample thickness leads to a

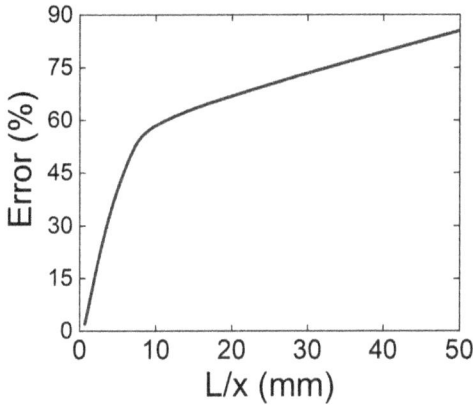

Fig. 5.4. Dependence of the n_2 estimation error on the value of ratio L/x. Reprinted from Ref. [39] with permission.

larger n_2 estimation error, and its decrease, on the contrary, to a smaller error. This assumption can be confirmed by numerical simulation for different wavelengths $\lambda_0 = 0.2$–0.6 mm and different sample thicknesses $L = 0.06$–4.5 mm. To compare the results for all values of the central wavelengths, the general parameter, L/x, is introduced; it is the ratio of the sample thickness, L, to the spatial size of the pulse, x. The dependence of the n_2 estimation error on ratio L/x is the same for all wavelengths and is shown in Fig. 5.4.

It can be seen that for broadband THz single-cycle pulsed radiation, the n_2 estimation error by the z-scan method grows with increasing ratio L/x. For a sample thickness corresponding to the L/x ratio of less than 10, a rapid increase in the n_2 estimation error is observed. For $L/x = 10$, the n_2 estimation error is more than 70%. Otherwise, when $L/x > 10$, the slope of the curve decreases, approaching values close to saturation (approximately 90%). Accordingly, the most accurate results are achieved in the case of the smallest value of the ratio. Since it is impossible to obtain an infinitely thin sample, the ratio $L/x \leq 1$ is the best option. In this case, the z-scan method has a negligible n_2 estimation error, which can be explained by the fact that nonlinear effects reveal themselves faster than dispersion effects.

5.3 Nonlinear Refractive Index Coefficient Estimation of Liquids in the THz Range

Let us discuss the experimental implementation of the z-scan method for THz radiation and the results of n_2 estimation of liquids in this spectral range, considering the features presented in the previous section.

Fig. 5.5(a) presents the experimental setup used in [47] to measure the coefficient of the nonlinear refractive index. The generation of high-intensity THz radiation occurs due to the optical rectification of femtosecond pulses in a lithium niobate crystal [48]. The pump radiation is a femtosecond laser system based on a regenerative amplifier with

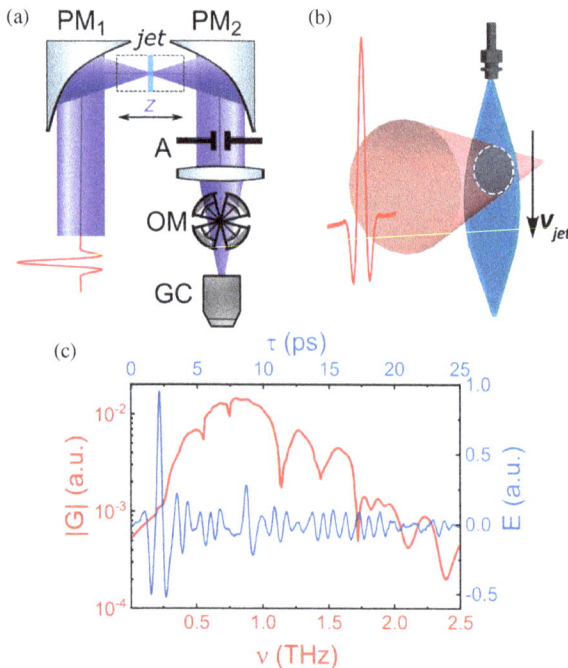

▲ Fig. 5.5. (a) Z-scan experimental setup for the nonlinear refractive index n_2 estimation: PM$_1$, PM$_2$ are parabolic mirrors, A is a closed aperture, OM is an optical modulator, and GC is a Golay cell. (b) Type and shape of an incident flat liquid jet from a nozzle. (c) Temporal and spectral structure of the THz pulse at the output of the THz generator. Reprinted from Ref. [28] with permission.

pulse duration 30 fs, pulse energy 2.2 mJ, repetition rate 1 kHz, and central wavelength 800 nm. The energy of the THz pulse at the generator output is 400–600 nJ, the pulse duration is 1.5 ps, and the spectral range is 0.1–2.5 THz (Fig. 5.5(c)–(d)). The intensity of the THz radiation is changed by reducing the intensity of the femtosecond pump or using crossed polarizers [49]. The polarization of the THz radiation at the generator output is vertical. Next, the pulsed THz radiation is focused and collimated by two parabolic mirrors (PM_1 and PM_2) with a focal length of 12.75 mm and forms a caustic with a diameter of about 1 mm. In this case, the peak intensity of the THz field is about 10^8 W/cm^2. Initially, the spatial size of the THz beam (diameter at $1/e^2$ level) at the output of the generator is about 25.4 mm. To avoid thermal heating of the medium, a liquid jet is used as a sample. Thus, each subsequent pulse interacts with a "new" medium (jet velocity 1 m/s, THz pulse repetition rate 1 kHz). Using a nozzle [50], the jet forms a quasi-plane-parallel plate (Fig. 5.5(b)). The jet displacement from −4 to 4 mm is restricted with the jet width of 8 mm and strong focusing of THz radiation. The jet has a thickness of 0.1 mm and is oriented along the normal to the incident radiation. With a duration of 1 ps, the spatial size of the pulse is 300 μm, i.e., the L/x ratio introduced in the previous section is ⅓, which is in line with the indicated recommendations. The THz radiation power is recorded by a Golay cell (GC). Aperture (A) is used for a closed aperture z-scan geometry. The THz field measurements is synchronized using a mechanical modulator (OM) located between the lens and the Golay cell.

In the described z-scan experiment, strong focusing is used with a beam diameter larger than the lens focus (Fig. 5.6), which is known to lead to the non-paraxial propagation nature after focusing. Even though the z-scan method implies paraxial radiation, it should be noted that the differences between the paraxial and non-paraxial modes for few-cycle pulses are insignificant [42]. To determine the influence of the non-paraxial propagation, numerical simulation can be carried out in the paraxial and non-paraxial approximations in the linear regime. Such a numerical simulation is demonstrated in [51]. To analyze the propagation

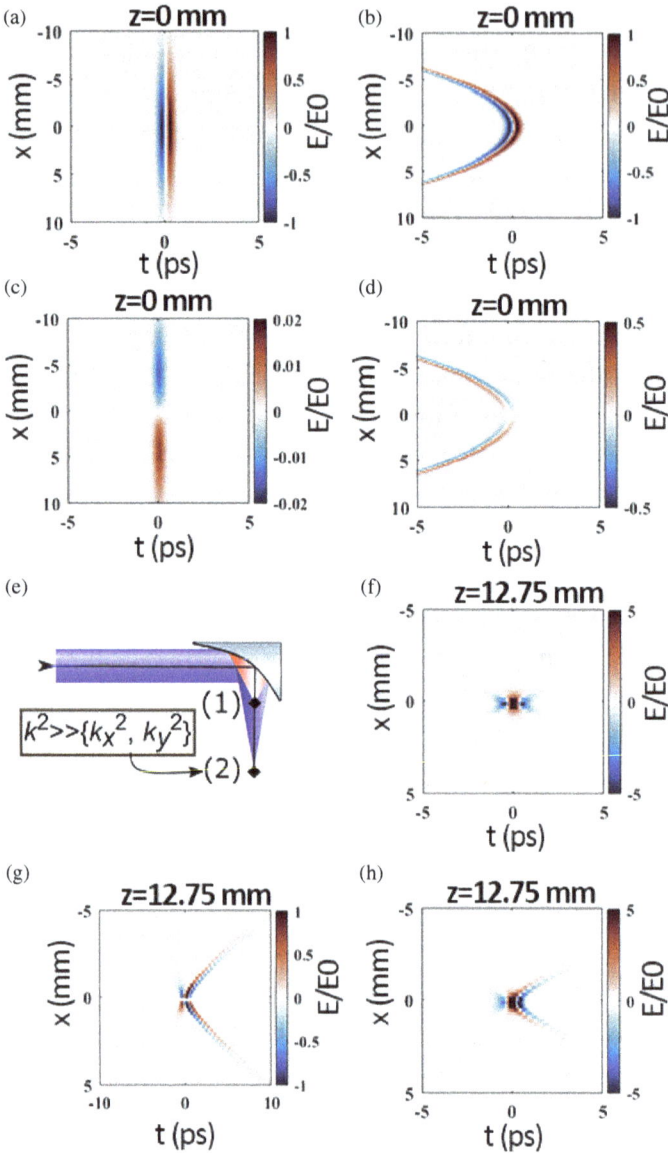

▲ Fig. 5.6. Spatiotemporal evolution of the THz wave electric field for non-paraxial and paraxial propagation from PM_1 mirror to the focal plane. (a–b) The electric field before the mirror and after applying the parabolic mirror defocusing for the transverse and (c–d) longitudinal components, respectively. (e) Illustration of non-paraxial (1) and paraxial (2) focusing. (g) The longitudinal component E_z after non-paraxial propagation. (f, h) Electric field profile at the focus for the transverse component E_x after paraxial and non-paraxial propagation, correspondingly. Reprinted from Ref. [28] with permission.

of the THz wavefront, a spectral approach is used [52]. Experimental linear focusing in the paraxial and non-paraxial approximations is estimated to determine their difference and applicability of the z-scan method for THz pulsed radiation.

Fig. 5.6 shows the spatiotemporal evolution of the electric field of the Gaussian THz wave. For the case of focusing, this beam profile is multiplied by the thin lens formula $exp\left[-\frac{ik}{2f}(x^2+y^2)\right]$ (Fig. 5.6(b)). The focal length of the parabolic mirror is $f=12.75$ mm. For the longitudinal component of the THz field, spatiotemporal changes are shown in Fig. 5.6(c)–(d). Strong focusing (at $f<x$) (Fig. 5.6(e)) provided by the numerical simulation results is comparable with experiment and leads to non-paraxial propagation. To calculate the non-paraxial propagation, the following coupled equations are used [52]:

$$
\begin{cases}
g_{x,y}\left(\omega,k_x,k_y,z\right)=C_{x,y}\left(\omega,k_x,k_y\right)e^{-i\sqrt{k^2-k_x^2-k_y^2}z}, \\
g_z\left(\omega,k_x,k_y,z\right)=\dfrac{k_xC_x\left(\omega,k_x,k_y\right)+k_yC_y\left(\omega,k_x,k_y\right)}{\sqrt{k^2-k_x^2-k_y^2}}e^{-i\sqrt{k^2-k_x^2-k_y^2}z}, \quad (5.7)
\end{cases}
$$

where $g_{x,y}$ and g_z are angular spectrum propagators for the transverse and the longitudinal field components respectively, constants C_x and C_y are determined from the boundary conditions, ω is central frequency of radiation, k_x and k_y are spatial frequency and k is wavenumber. Calculations of longitudinal component E_z and transverse component E_x for the non-paraxial case are shown in Fig. 5.6(g) and (h), respectively.

The spatio-spectral evolution of the electric field for the paraxial approximation in this case is calculated using Eq. (5.7):

$$
g_{x,y}\left(\omega,k_x,k_y,z\right)=C_{x,y}\left(\omega,k_x,k_y\right)e^{-ikz\left(1-\frac{k_x^2+k_y^2}{2k^2}\right)}. \qquad (5.8)
$$

Fig. 5.6(e) demonstrates the paraxial calculation of spatiotemporal evolution of the transverse component E_x.

To compare the two approximations, the ratio of the longitudinal component to the transverse component is calculated:

$$W = \frac{\int_{-\infty}^{+\infty}\int_{-\infty}^{+\infty}\int_{-\infty}^{+\infty} \left|E_z(x,y,t)\right|^2 dxdydt}{\int_{-\infty}^{+\infty}\int_{-\infty}^{+\infty}\int_{-\infty}^{+\infty} \left|E_x(x,y,t)\right|^2 dxdydt} = 7.6\%. \tag{5.9}$$

A theoretical analysis shows that the ratio of the longitudinal component to the transverse component calculated for parameters close to the experimental is 7.6%. Therefore, for the analysis of the experiment, the non-paraxial nature of beam propagation can be neglected.

Finally, let us discuss the results of experimental measurements. Fig. 5.7 shows a comparison of the z-scan curves for a jet of water (Fig. 5.7(a)), α-pinene (Fig. 5.7(b)), ethanol (Fig. 5.7(c)) and 2-propanol (Fig. 5.7(d)). According to Eqs. (5.4) and (5.5), the solid lines in the figure represent the analytical z-scan curves for monochromatic radiation with a wavelength of $\lambda_0 = 0.4$ mm ($v_0 = 0.75$ THz). The following media parameters are used for the analytical curves: absorption coefficient (water $\alpha = 100$ cm^{-1} α-pinene $\alpha = 10$ cm^{-1}, ethanol $\alpha = 60$ cm^{-1}, 2-propanol $\alpha = 40$ cm^{-1}), sample length $L = 0.1$ mm, THz pulse duration $\tau_0 = 1$ps at 1/e energy level of the temporal structure, central radiation frequency $v_0 = 0.75$ THz ($\lambda_0 = 0.4$ mm), waist radius $w_0 = 0.5$ mm, aperture radius $r_a = 1.5$ mm, initial THz beam radius $w_a = 12.5$ mm, and intensity in the focal point $I_0 = 0.5 \times 10^8$ W/cm^2. The nonlinear refractive index coefficient is varied to obtain the best agreement between the experimental and analytical curves. It should be noted that the linear transmission of the aperture is 2%; this value makes it possible to maximize the sensitivity of the measurement method but reduces the signal-to-noise ratio. This is because the z-scan method is sensitive to changes in the intensity of the Gaussian beam profile peak. Reducing the aperture leads to a decrease in the average transmission, without changing the peak-to valley ratio, which gives a gain in relative terms.

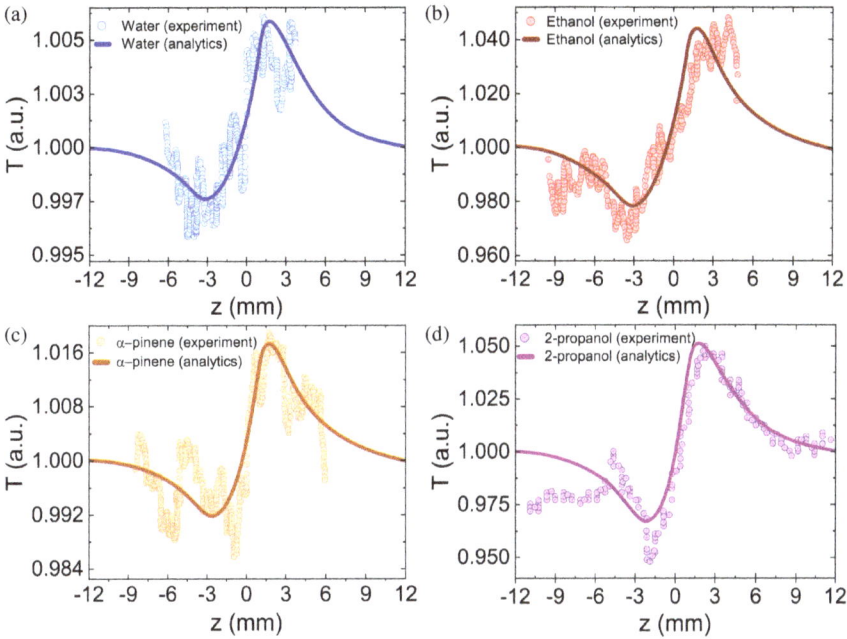

▲ Fig. 5.7. Comparison of experimental and analytical dependencies of normalized transmission T through the aperture on the jet position z for (a) water, (b) ethanol, (c) α-pinene, and (d) 2-propanol. (a), (b) and (c) are reprinted with permission from Tcypkin A., Zhukova M., Melnik M., Vorontsova I., Kulya M., Putilin S., ... & Boyd R. W. (2021). Giant third-order nonlinear response of liquids at terahertz frequencies, Physical Review Applied, 15(5), p. 054009.

As seen in Fig. 5.7, the experimental z-scan curve for broadband THz radiation agrees well with the analytical z-scan curve for monochromatic radiation. The deviation of the maxima and minima positions can be explained by the fact that the analytics is intended for monochrome radiation, while broadband radiation is used in the experiment. However, the peak-to-valley ratios are in good agreement with each other.

The asymmetry of the z-scan curve is caused by the presence of nonlinear absorption or bleaching. To estimate the nonlinear refractive index coefficient, two measurements are used – with open and closed apertures (Fig. 5.8(a)–(b) for water). To correctly determine the nonlinear refractive index coefficient,

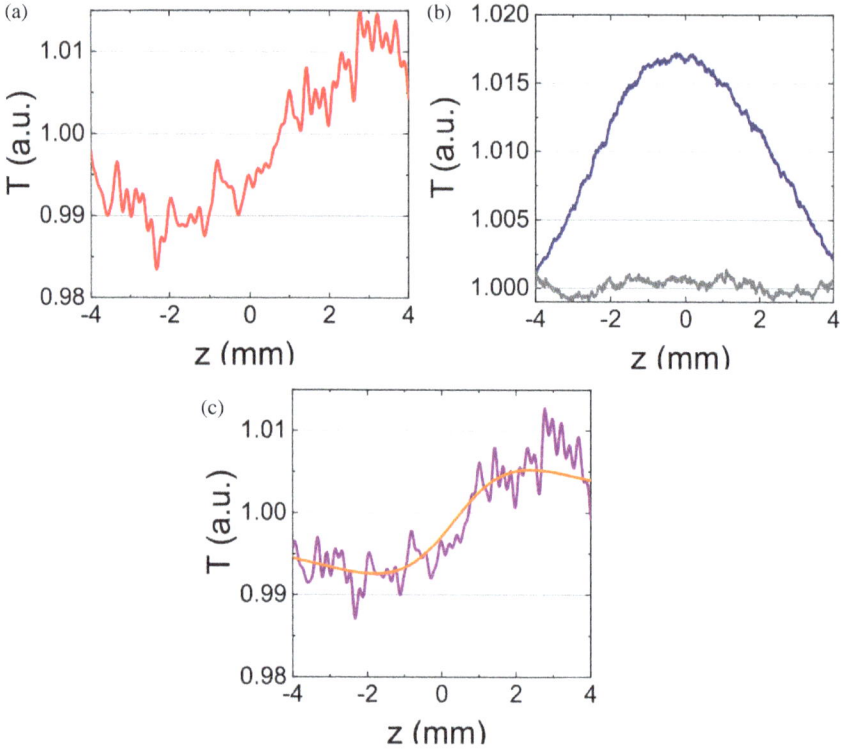

▲ Fig. 5.8. Procedure for the analysis of experimental data in the case of water. (a) Real experimental curves for closed and (b) open apertures. (c) Curve obtained by dividing the closed aperture curve by the open aperture curve (purple curve) showing symmetry between peak and valley, and analytical z-scan curve for monochromatic radiation with a wavelength of 0.4 mm (orange). Reprinted from Ref. [28] with permission.

it is necessary to divide the results of the experiment with a closed aperture by an open aperture (Fig. 5.8(c)). It is these curves that are used to calculate n_2 in experiments for the IR and optical ranges of the spectrum.

To estimate n_2 from the experimental data, standard z-scan equations are used [35]:

$$n_2 = \frac{\Delta T}{0.406 I_{in}} \cdot \frac{\sqrt{2}\lambda_0}{\left(2\pi L_\alpha \left(1-S\right)^{0.25}\right)}, \qquad (5.10)$$

where ΔT (Fig. 5.7) is the difference between the maximum and minimum transmission, S is the linear transmission of the aperture, L is the sample thickness, $L_\alpha = \alpha^{-1}[1 - exp(-\alpha L)]$ is the effective interaction length, α is the absorption coefficient, λ_0 is the wavelength, and I_{in} is the input radiation intensity. The radiation wavelength is chosen as $\lambda_0 = 0.4$ mm ($\nu_0 = 0.75$ THz), which corresponds to the maximum of the THz radiation generation spectrum (Fig. 5.5(c)).

Thus, the nonlinear refractive index coefficients calculated by Eq. (5.10) using experimental data for different liquids have the following values: water -7×10^{-10} cm^2/W, α-pinene -3×10^{-9} cm^2/W, ethanol -6×10^{-9} cm^2/W, and 2-propanol -2×10^{-9} cm^2/W. In the THz range these values are 4–6 orders higher compared to the visible and IR ranges of the spectrum [53–55]. The results of the experiments discussed raise a logical question about the nature of such a high nonlinearity of the refractive index of the studied materials. It is this question that we discuss in the next section.

5.4 The Giant Nonlinearity Nature of Refractive Index for Liquids in the THz Range

It was theoretically predicted in [22] that anharmonic vibrations of atoms in the molecules of a dielectric medium can be the dominant low-inertia mechanism of the refractive index nonlinearity in the field of THz waves, and such a nonlinearity is gigantic. The works discussed in the previous section experimentally found that the coefficient of the nonlinear refractive index of water and some other liquids in the THz spectral range (0.1–2 THz) is a million times higher than in the visible and near-IR ranges. In this section, we will show that the measured values of the nonlinear refractive index coefficient of liquids correspond to the theory of anharmonic molecular vibrations of atoms in the molecules of these liquids [28].

The mathematical model of atom oscillations in a molecule of a dielectric medium in the field of linearly polarized radiation of the far-IR

range in the simplest form can be represented by an anharmonic oscillator equation [22, 28]:

$$\frac{\partial^2 x}{\partial t^2} + \gamma \frac{\partial x}{\partial t} + \omega_0^2 x + a x^2 + b x^3 = \alpha E, \tag{5.11}$$

where x is the deviation of atom in the molecule from the equilibrium position, t is the time, γ is the damping coefficient of the molecular oscillator, ω_0 is the frequency of free oscillations, a and b are the coefficients characterizing the quadratic and cubic anharmonicity of molecular oscillations, respectively, α is the polarization coefficient of the medium, and E is the electromagnetic field of the THz pulse.

In education, nonlinear phenomena tend to be regarded in the framework of the classical theory of dispersion of intense radiation, which is based on the equations of anharmonic vibrations of the structural elements of matter. Such a simple and clear approach is used in almost every textbook on nonlinear optics. But the applicability of this approach to practical estimation of the characteristics of optical media nonlinearity, for example, nonlinear refractive index coefficient, is not obvious. Therefore, numerous publications of the previous century did not apply the classical theory of intense light dispersion to theoretical analysis and attempts to calculate n_2 of the electronic nature for transparent optical materials (see, for example, [56]). The very anharmonicity of an electronic oscillator can be described by Eq. (5.11), but it is impossible to estimate coefficients a and b that characterize this anharmonicity. It is a completely different case with the analysis of the nonlinear response of an oscillatory nature. The model of anharmonic vibrations of a molecular oscillator (Eq. (5.11)) in the absence of an external electromagnetic field is also widely used in solid state physics to describe the thermal expansion of matter [57]. In this model, the coefficient of thermal expansion of matter is represented by the following equation:

$$\alpha_T = -\frac{a k_B}{m \omega_0^4 a_1}, \tag{5.12}$$

where a is the coefficient describing in Eq. (5.11) the quadratic nonlinear response of the molecule, k_B is the Boltzmann constant, m is the reduced mass of the vibrational mode, and a_1 is the lattice constant (for solids) or the diameter of the molecule (for liquids). Therefore, as shown in [22], the remaining coefficients of the anharmonic molecular oscillator model (Eq. (5.11)) and anharmonicity coefficient a can be expressed in terms of other medium characteristics already known from literature, for example, the thermal expansion coefficient of matter, linear refractive index of the medium, and frequency of atomic stretching vibrations in molecules. The expressions are given below. To begin with, let us consider a characteristic derivation of the nonlinear dynamic equations of medium polarization using the molecular response model (Eq. (5.11)).

Assuming the nonlinearity of molecular vibrations to be small, the solution of Eq. (5.11) is the following:

$$x = x^{(1)} + x^{(2)} + x^{(3)}, \tag{5.13}$$

where $x^{(1)}$, $x^{(2)}$ and $x^{(3)}$ are functionals proportional to the first, second and third order of the electrical field, respectively, and for which the inequality implemented in practice is almost always satisfied:

$$x^{(1)} >> x^{(2)}, x^{(3)}. \tag{5.14}$$

Thus, Eq. (5.11) can be represented as coupled equations [22]

$$\left\{ \begin{aligned} &\frac{\partial^2 x^{(1)}}{\partial t^2} + \gamma \frac{\partial x^{(1)}}{\partial t} + \omega_0^2 x^{(1)} = \alpha E, \\ &\frac{\partial^2 x^{(2)}}{\partial t^2} + \gamma \frac{\partial x^{(2)}}{\partial t} + \omega_0^2 x^{(2)} + a\left(x^{(1)}\right)^2 = 0, \\ &\frac{\partial^2 x^{(3)}}{\partial t^2} + \gamma \frac{\partial x^{(3)}}{\partial t} + \omega_0^2 x^{(3)} + 2ax^{(1)}x^{(2)} + b\left(x^{(1)}\right)^3 = 0. \end{aligned} \right. \tag{5.15}$$

In liquids, such a macroscopic characteristic of a medium as its polarization of oscillatory nature can be considered as

$$P = Nq\langle x \rangle, \tag{5.16}$$

where N is the concentration of molecules, q is the charge of atoms, and the angle brackets mean the averaging of the deviations from the equilibrium position of atoms in molecules over a unit volume of the medium. Then, for liquids, which are isotropic under normal conditions with the response of an individual molecule (Eq. (5.15)), the time evolution of the polarization of the medium (Eq. (5.16)) in an arbitrary optical field is characterized by the following equations [58]:

$$\begin{cases} \dfrac{\partial^2 P_{lin}}{\partial t^2} + \gamma \dfrac{\partial P_{lin}}{\partial t} + \omega_0^2 P_{lin} = Nq\alpha E, \\[2mm] \dfrac{\partial^2 P_{nl}}{\partial t^2} + \gamma \dfrac{\partial P_{nl}}{\partial t} + \omega_0^2 P_{nl} = R P_{lin} - \dfrac{b}{(Nq)^2} P_{lin}^3, \\[2mm] \dfrac{\partial^2 R}{\partial t^2} + \gamma \dfrac{\partial R}{\partial t} + \omega_0^2 R = \dfrac{2a^2}{(Nq)^2} P_{lin}^2. \end{cases} \tag{5.17}$$

Here, the first equation describes the time evolution of the linear part of the polarization response of oscillatory nature, P_{lin}, and the second and third equations are parametrically coupled and describe the cubic nonlinear polarization response P_{nl} of the medium. The parameter R is defined as $R = -2ax^{(2)}$.

The quadratic polarization response of the medium vanishes due to its symmetry when individual molecular quadratic responses (Eq. (5.15)) are averaged over their large ensemble in a unit volume in an isotropic medium (Fig. 5.9).

The classical nonlinear theory of radiation dispersion in isotropic media is often used in theoretical studies based on the constitutive equation

$$\frac{\partial^2 P}{\partial t^2} + \gamma \frac{\partial P}{\partial t} + \omega_0^2 P + \frac{b}{(Nq)^2} P^3 = qN\alpha E, \tag{5.18}$$

to which Eq. (5.17) is reduced by neglecting the quadratic nonlinearity of vibrations of an individual molecule. As demonstrated in [43], Eq. (5.18)

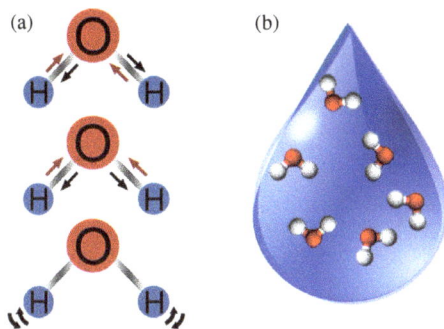

▲ Fig. 5.9.   Explanation of the absence of a quadratic response nonlinearity of an isotropic medium in the presence of a quadratic response nonlinearity of its molecules by the example of water. (a) Types of vibrations in the H_2O molecule. (b) Different orientations of molecules in a water drop (water jet in the experiment). Reprinted from Ref. [28] with permission.

does not qualitatively correspond to quantum theory; for example, it does not predict a well-known effect in nonlinear optics such as an increase in the nonlinear refractive index at two-photon resonance. At the same time, as shown in [58], Eq. (5.17) describes this effect in detail. It was heuristically shown [59] that the classical model of light-matter interaction, which allows us to obtain the nonlinear refractive index dispersion, must include not one but at least two parametrically coupled nonlinear oscillators to be consistent with the results of quantum calculations. Thus, in an isotropic material, the quadratic nonlinearity of the response of an individual molecule due to averaging does not excite a quadratic response of the volume of the medium. Nevertheless, the quadratic nonlinearity of the molecular oscillator contributes to the cubic nonlinearity of the macroscopic polarization, which is described by the third Eq. (5.17).

In accordance with [58], the constitutive Eqs. (5.17) obtained for a nonlinear isotropic medium can be transformed for the special case of quasi-monochromatic pulsed THz radiation:

$$E = \frac{1}{2}\varepsilon_\omega(t)\exp(i\omega t) + c.c.,\qquad (5.19)$$

where $\varepsilon_\omega(t)$ is the complex amplitude of the electromagnetic field with the carrier frequency ω. Eqs. (5.17) assume the following:

$$P_{lin} = \frac{1}{2}P_\omega^{lin}(t)exp(i\omega t)+c.c.,\tag{5.20}$$

$$P_{nl} = \frac{1}{2}\left(P_\omega^{nl}(t)exp(i\omega t)+c.c.+P_{3\omega}^{nl}(t)exp(i3\omega t)+c.c.\right),\tag{5.21}$$

$$R = R_0(t)+\frac{1}{2}\left(R_{2\omega}(t)exp(i\omega t)+c.c.\right),\tag{5.22}$$

where $P_\omega^{lin}(t)$, $P_\omega^{nl}(t)$ are the complex amplitudes of the linear and nonlinear parts of the polarization response at the carrier frequency of the THz quasi-monochromatic pulse, $P_{3\omega}^{nl}(t)$ is the nonlinear polarization at tripled frequency, and $R_0(t)$ and $R_{2\omega}(t)$ are oscillations of parametrically coupled oscillators at "zero" and doubled frequencies, respectively.

Eqs. (5.17) for the polarization response of the medium at the frequency of exciting field ω with new variables (Eqs. (5.20)–(5.22)) takes the form:

$$\frac{\partial^2 P_\omega^{lin}}{\partial t^2}+(\gamma+2i\omega)\frac{\partial P_\omega^{lin}}{\partial t}+\left(\omega_0^2-\omega^2+i\gamma\omega\right)P_\omega^{lin}=Nq\alpha\varepsilon_\omega,\tag{5.23}$$

$$\frac{\partial^2 P_\omega^{nl}}{\partial t^2}+(\gamma+2i\omega)\frac{\partial P_\omega^{nl}}{\partial t}+\left(\omega_0^2-\omega^2+i\gamma\omega\right)P_\omega^{nl}=R_0 P_\omega^{lin}$$

$$+\frac{1}{2}R_{2\omega}\left(P_\omega^{lin}\right)^*-\frac{3}{4}\frac{b}{\left(Nq\right)^2}\left|P_\omega^{lin}\right|^2 P_\omega^{lin},$$

$$\frac{\partial^2 R_{2\omega}}{\partial t^2}+(\gamma+4i\omega)\frac{\partial R_{2\omega}}{\partial t}+\left(\omega_0^2-4\omega^2+2i\gamma\omega\right)R_{2\omega}=\frac{a^2}{\left(Nq\right)^2}\left(P_\omega^{lin}\right)^2,$$

$$\frac{\partial^2 R_0}{\partial t^2}+\gamma\frac{\partial R_0}{\partial t}+\omega_0^2 R_0=\frac{a^2}{\left(Nq\right)^2}\left|P_\omega^{lin}\right|^2.$$

In [58] the settling times are obtained for a nonlinear response of oscillatory nature for α-pinene, water, and silicon dioxide as media with a particularly high oscillatory nonlinearity of the refraction index. The estimates are based on equations of nonlinear polarization dynamics of an isotropic medium (Eq. (5.23)) as well as on Eqs. (5.17). For these materials the time constants of the resonant oscillatory mechanism inertia of their nonlinearity in the case of radiation in the THz range are hundreds of femtoseconds, and in the case of nonresonant interaction they decrease to ten femtoseconds or less.

Theoretical substantiation for the low inertia of the nonlinear refractive index of oscillatory nature is fully consistent with the experiments on measuring the nonlinear refractive index coefficient in fast flowing liquid jets, which are described in the previous section. Thus, the anharmonicity of molecular vibrations (Eq. (5.11)) describes both the slow thermal expansion of media and the rapid onset of refractive index nonlinearity in the THz spectral range.

For long pulses, for which the inequality

$$\tau_{pulse} \gg \omega^{-1}, \gamma^{-1}, \omega_0^{-1} \tag{5.24}$$

is satisfied, the nonlinear polarization response of the medium, as follows from Eqs. (5.23), can be considered inertialess, and it takes the form [22]

$$P_\omega^{nl} = \frac{3}{4} \chi^{(3)} |\varepsilon_\omega|^2 \varepsilon_\omega, \tag{5.25}$$

where cubic nonlinear susceptibility can be described as [56]

$$\chi_\omega^{(3)} = \frac{1}{3} \frac{qN\alpha^3}{\left(\omega_0^2 - \omega^2 + i\gamma\omega\right)^2} \frac{1}{\left(\omega_0^2 - \omega^2\right)^2 + \left(\gamma\omega\right)^2} \left[2a^2 \frac{3\omega_0^2 - 8\omega^2 + 4i\gamma\omega}{\omega_0^2 \left(\omega_0^2 - 4\omega^2 + 2i\gamma\omega\right)} - 3b \right] \tag{5.26}$$

In an isotropic medium with a cubic polarization response in a strong field of intense optical radiation, E, the refractive index of the optical medium takes the form:

$$n(\omega) = n_0(\omega) + \frac{1}{2} n_2(\omega) |E|^2, \tag{5.27}$$

where the nonlinear refractive index coefficient of the medium, n_2, is related to the cubic nonlinear susceptibility (Eq. (5.26)) by the relation

$$n_2 = \frac{3\pi}{n_0}\chi^{(3)}. \tag{5.28}$$

Then, in the framework of the classical theory of intense light dispersion, n_2 of an isotropic medium can be written as [56]

$$n_2 = \frac{\pi}{n_0}\frac{qN\alpha^3}{\left(\omega_0^2-\omega^2+i\gamma\omega\right)^2}\frac{1}{\left(\omega_0^2-\omega^2\right)^2+\left(\gamma\omega\right)^2}\left[2a^2\frac{3\omega_0^2-8\omega^2+4i\gamma\omega}{\omega_0^2\left(\omega_0^2-4\omega^2+2i\gamma\omega\right)}-3b\right] \tag{5.29}$$

The key idea of [22] is to obtain expressions for other physical parameters along with the coefficient of the nonlinear refractive index (Eq. (5.29)) within the framework of a simple model of molecular vibrations (Eq. (5.11)). Then, these expressions allow us to determine the dependence of the required nonlinear coefficient on these known physical parameters.

Dolgaleva et al. [22] showed that if the cubic nonlinearity of polarization response for an optical medium in the THz range is of oscillatory nature, the factors in Eq. (5.29) are related to the known optical, spectral, and thermal characteristics of the medium by:

$$\alpha = \frac{\omega_0^2}{4\pi qN}\left(n_{0,v}^2-1\right), \tag{5.30}$$

$$a = -\frac{m\omega_0^4 a_1}{k_B}\alpha_T, \tag{5.31}$$

$$b = \frac{6\pi q^2 N\omega_0}{\left(n_{0,v}^2-1\right)\hbar}, \tag{5.32}$$

where $n_{0,v}$ is a part of the linear refractive index of a medium of oscillatory nature at $\omega \ll \omega_0$, and $\hbar$ is the reduced Planck constant.

Eq. (5.29) is obtained in the CGS system. To convert it to SI, it is necessary to consider the relationship between the value of the nonlinear refractive index coefficient in the CGS system (Eq. (5.27)) and in the SI system (Eq. (5.1)):

$$n_2\left(\frac{cM^2}{BT}\right) = 4.2 \cdot 10^{-4}\, \frac{n_2\left(C\Gamma C\right)}{n_0}.\tag{5.33}$$

Thus, n_2 of the medium (in SI), taking into account expressions for known values (Eqs. (5.30)–(5.32)), can be represented as [22]:

$$n_2 = \frac{3}{32n_0}\,\frac{m^2\omega_0^4 a_i^2}{\pi^2 q^2 N^2 k_B^2}\,\alpha_T^2\left(n_{0,v}^2 - 1\right)^3 - \frac{9}{32\pi N n_0 \hbar\omega_0}\left(n_{0,v}^2 - 1\right)^2,\tag{5.34}$$

where a_i is the unit cell size (for solids) or molecular diameter (for liquids), m is the reduced mass of the vibrational mode, ω_0 is the fundamental frequency of vibrations, α_T is the thermal expansion coefficient, q is the effective charge of the chemical bond, N is the density of molecular oscillators, defined as the ratio of the specific gravity to the total mass of the molecule, and $n_{0,v}$ corresponds to the vibrational contribution to the low-frequency refractive index

$$n_{0,v} = \sqrt{1 + n_0^2 - n_{el}^2},\tag{5.35}$$

where n_0 is the linear refractive index of the medium in the THz region, and n_{el} is the part of the refractive index due to the electronic contribution.

Eq. (5.34) is written in the approximation $\omega \ll \omega_0$, which is applicable to the media we are considering. For example, water has $\omega_0 \sim 100$ THz ($\lambda = 3\ \mu m$) [60]. In the experiments presented in the previous chapter, the value of $\omega/2\pi$ is approximately 0.75 THz (Fig. 5.5(c)), which is much less than the value of $\omega_0/2\pi$: 15.9 THz.

To estimate the nonlinear refractive index coefficient of liquids using Eq. (5.34), the values given in Table 5.1 were used, where ω_0 is the fundamental frequency of molecular vibrations, n_{el} is the refractive index

▼ Table 5.1. Parameters used to calculate the coefficient of the nonlinear refractive index n_2 for the THz range.

Liquid	$\omega/2\pi$, THz	n_0	n_{el}	$a_l \times$ 10^{-7} cm	$\alpha_T \times$ 10^{-4} °C^{-1}	S	n_2^{IR}, cm²/W	n_2^{theory}, cm²/W	$n_2^{experiment}$, cm²/W
Water	100 [60]	2.3 [61]	1.3290 [60]	0.28 [62]	2 [63]	1	4.1×10^{-16} [53]	5×10^{-10}	$5-9 \times 10^{-10}$ 7.8×10^{-10} [30] 6.5×10^{-10} [34] 7.2×10^{-10} [33] 2.3×10^{-10} [64]
2-propanol	89.2 [65]	1.45 [66]	1.3786 [67]	0.6 [68]	11.7 [69]	0.785	1.38×10^{-16} [67]	2×10^{-9}	1.7×10^{-9}
Ethanol	115 [70]	1.55 [71, 72]	1.3575 [73]	0.31	14.9 [74]	0.79 [66]	7.7×10^{-16} [53]	9×10^{-9}	$4-8 \times 10^{-9}$
α-pinene	90 [76]	1.51	1.4663 [74]	1.2	9 [77]	0.86	1.56×10^{-15} [54]	1×10^{-9}	$1-5 \times 10^{-9}$

of the medium in the near-IR range (800 nm), n_0 is the refractive index in the range of 0.1–1 THz, a_l is the distance between atoms in a molecule (molecular diameter), m is the reduced mass of the vibrational mode, α_T is the thermal expansion coefficient, S is the relative density, q is the effective charge of the chemical bond, in this case the electron charge, and N is the number of molecular oscillators per unit volume, which is calculated as the ratio of the specific gravity to the total mass of the molecule. For example, in the case of water, the number of vibrational units is 1, the mass of the H_2O molecule is $(1 \times 2 + 16)$ times the atomic mass unit (1.67×10^{-24}). The result is $N = 3.3 \times 10^{22}$ per 1 cm³. Eq. (5.34) was obtained in the framework of a theoretical approach that was originally developed for solids. To apply Eq. (5.34) to liquids, the approach must be refined. In particular, redefine the mass of the vibrational mode as the reduced mass of the vibrational mode of one bond. For example, in the case of water and ethanol, the O–H bond is used for calculations, and in α-pinene C–H. The length of these bonds is approximately the same and corresponds to an intermediate value between the length of a single bond (0.154 nm) and a double bond (0.134 nm).

▼ Table 5.2. Nonlinear refractive index n_2 of various media in the THz range.

Media	n_2, cm²/W
Water vapor	6×10^{-11} [30]
SiO$_2$	3.5×10^{-12} [78] 4.4×10^{-12} [22]
ZnSe	2.5×10^{-11} [45]
GaP	1.2×10^{-13} [79]
LiNbO$_3$	4×10^{-11} [80]

As seen from the table, the experimentally obtained z-scan curves for broadband THz radiation and the calculation of the value of n_2 based on them are in good agreement with the theoretical estimation.

Table 5.2 shows the results of n_2 estimation of other materials in the THz frequency range. It can be seen that the value of n_2 of liquid media is higher than provided values.

Strictly speaking, not only the oscillatory mechanism of the nonlinearity can contribute to the nonlinearity of the refractive index (Eq. (5.1)) in liquids in the THz frequency range. Let us compare the results of theoretical estimation of n_2 for water with the contribution to this value of thermal nonlinearity and molecular orientation.

To estimate the contribution of thermal nonlinearity to the refractive index, the following formula can be used [81]:

$$n_2^{(th)} = \left(\frac{dn}{dT} \right) \frac{\alpha R^2}{\kappa}.$$

(5.36)

Here, dn/dT is the temperature coefficient of the refractive index, α is the linear absorption coefficient of the material, k is the thermal conductivity, and R is the radius of the laser beam.

For our case, the values are $dn/dT = -9.5 \times 10^{-5}$ K$^{\circ-1}$ [82], $\alpha = 200$ cm^{-1} [77], $k = 0.595$ W/(m $\times$ K$^{\circ}$) [82]; for $R = 0.5$ mm, the calculated coefficient is $n_2^{(th)} = 8 \times 10^{-3}$ cm^2/W.

In turn, the orientational contribution to the refractive index nonlinearity is defined as [81]:

$$\bar{n}_2 = \frac{N}{45 n_0} \left(\frac{n_0^2 + 2}{3} \right)^4 \frac{(\alpha_3 - \alpha_1)^2}{kT}, \tag{5.37}$$

where N is the number density of the molecules, α_3 denotes the polarizability experienced by an optical field parallel to its axis of symmetry, and α_1 denotes the polarizability experienced by a field perpendicular to its axis of symmetry.

It is known that $N = 3 \times 10^{22}$ cm^{-1}, $n_0 = 1.33$, $\alpha_3 = 1.86$ Å^3 [83], and $\alpha_1 = 1.63$ Å^3 [83]. For temperature $T = 293.15$ K$^{\circ}$, $\bar{n}_2 = 1.6 \times 10^{-15}$ esu, and $n_2^{(or)} = 6.9 \times 10^{-18}$ cm^2/W.

Therefore, the contribution of thermal nonlinearity is $n_2^{(th)} = 8 \times 10^{-3}$ cm^2/W, and the orientational contribution to the refractive index nonlinearity is $n_2^{(or)} = 6.9 \times 10^{-18}$ cm^2/W. The electronically induced change in the refractive index [84] is $n_2^{(el)} = -9 \times 10^{-15}$ cm^2/W. The latter two parameters are negligible due to their small values. Although the thermal effects are quite high relative to the calculated value, their influence on the change in the refractive index can be neglected.

The estimated response time, τ_r, of the thermal nonlinearity for the described case can be calculated as follows:

$$\tau_r \approx \frac{(\rho_0 C) R^2}{\kappa}. \tag{5.38}$$

Using the given equation, $\tau_r = 1.75$ s, while the plane-parallel water jet used in the experiments had a velocity of 10^3 mm/s. This means that the geometric region of interaction between the jet and light was constantly changing, leaving no time for the appearance of thermal effects. The thermal contribution is approximately 10^{-14} considering its linear dependence on the time of interaction with the medium, which in the case of a jet is tens of ps (10^{-12} s).

Based on the above calculations, it can be concluded that the main mechanism of low-inertia nonlinearity in the THz frequency range has oscillatory nature.

5.5 Conclusion

This chapter has discussed the features of recently discovered giant low-inertia nonlinearity of refractive index for liquids in the THz range. The limits of applicability were determined for the method for broadband pulsed THz radiation with a small number of oscillations. The experimental setup for measuring low-inertia n_2 of liquids by this method was described. It was shown that the methodological error in the measurement of n_2 decreases with an increase in the number of oscillations and practically vanishes for pulses with three or more field oscillations. For few-cycle THz pulses the decrease in the measurement error can be obtained by reducing the ratio of the longitudinal pulse size to the thickness of the medium under study. The use of liquid jets made it possible to distinguish the low-inertia nonlinearity of the refractive index in the n_2 measurements. This chapter also shows that the simple classical model of molecular oscillators well describes the experimental results.

References

1. Zhang X.-C. & Xu J. (2010). *Introduction to THz Wave Photonics.* Springer.
2. Samanta D. *et al.* (2022). Trends in Terahertz Biomedical Applications, in *Lecture Notes in Electrical Engineering*, Springer, pp. 285–299.
3. Krügener K. *et al.* (2020). Terahertz inspection of buildings and architectural art, Applied Sciences (Switzerland), 10, pp. 5166.
4. Liu X. *et al.* (2020). Formation of gigahertz pulse train by chirped terahertz pulses interference, Scientific Reports, 10, pp. 1–7.

5. Song H. J. (2021). Terahertz wireless communications, IEEE Microwave Magazine, 22, pp. 88–99.

6. Devi N., Dash J., Ray S. & Pesala B. (2017). Non-invasive characterization of carbon fiber reinforced polymer composites using continuous wave terahertz system, 9th International Conference on Trends in Industrial Measurement and Automation, IEEE, pp. 1–4.

7. Grachev Y. V. *et al.* (2018). Wireless data transmission method using pulsed THz sliced spectral supercontinuum, IEEE Photonics Technology Letters, 30, pp. 103–106.

8. Dong J., Locquet A., Melis M. & Citrin, D. S. (2017). Global mapping of stratigraphy of an old-master painting using sparsity-based terahertz reflectometry, Scientific Reports, 7, pp. 1–12.

9. Fülöp J. A., Tzortzakis S. & Kampfrath T. (2020). Laser-driven strong-field terahertz sources, Advanced Optical Materials, 8, pp. 1900681.

10. Kim K.-Y., Glownia J. H., Taylor A. J. & Rodriguez G. (2007). Terahertz emission from ultrafast ionizing air in symmetry-broken laser fields, Optics Express, 15, pp. 4577.

11. Bartel T., Reimann K., Woerner M. & Elsaesser T. (2005). Generation of THz transients with high electric-field amplitudes, Optics InfoBase Conference Papers, 30, pp. 2805–2807.

12. Hirori H., Doi A., Blanchard F. & Tanaka K. (2011). Single-cycle terahertz pulses with amplitudes exceeding 1 MV/cm generated by optical rectification in LiNbO3, Applied Physics Letters, 98, pp. 091106.

13. Hebling J., Yeh K. lo, Hoffmann M. C. & Nelson K. A. (2008). High-power THz generation, THz nonlinear optics, and THz nonlinear spectroscopy, IEEE Journal on Selected Topics in Quantum Electronics, 14, pp. 345–353.

14. Shalaby M. & Hauri C. P. (2015). Demonstration of a low-frequency three-dimensional terahertz bullet with extreme brightness, Nature Communications, 6, pp. 1–8.

15. Sell A., Leitenstorfer A. & Huber R. (2008). Phase-locked generation and field-resolved detection of widely tunable terahertz pulses with amplitudes exceeding 100 MV/cm, Optics Letters, 33, pp. 2767.

16. Junginger F. *et al.* (2010). Single-cycle multiterahertz transients with peak fields above 10 MV/cm, Optics Letters, 35, pp. 2645.

17. Chai X. *et al.* (2018). Subcycle terahertz nonlinear optics, Physical Review Letters, 121, pp. 143901.

18. Hafez H. A. *et al.* (2020). Terahertz nonlinear optics of graphene: from saturable absorption to high-harmonics generation, Advanced Optical Materials, 8, pp. 1900771.
19. Zhang Y., Li K. & Zhao H. (2021). Intense terahertz radiation: generation and application, Frontiers of Optoelectronics, 14, pp. 4–36.
20. Jencson J. E. *et al.* (2014). Focus on nonlinear terahertz studies, New Journal of Physics, 16, pp. 045016.
21. Tcypkin A. *et al.* (2021). Surprising nonlinear optics of pulsed terahertz radiation, International Conference on Advanced Laser Technologies (ALT), 21, pp. 179–179.
22. Dolgaleva K., Materikina D. V., Boyd R. W. & Kozlov S. A. (2015). Prediction of an extremely large nonlinear refractive index for crystals at terahertz frequencies, Physical Review A, 92, pp. 023809.
23. Turchinovich D., Hvam J. M. & Hoffmann M. C. (2012). Self-phase modulation of a single-cycle terahertz pulse by nonlinear free-carrier response in a semiconductor, Physical Review B, 85, pp. 201304.
24. Korpa C. L., Tóth G. & Hebling J. (2016). Interplay of diffraction and nonlinear effects in the propagation of ultrashort pulses, Journal of Physics B: Atomic, Molecular and Optical Physics, 49, pp. 035401.
25. Zhukova M., Melnik M., Vorontsova I., Tcypkin A. & Kozlov S. (2020). Estimations of low-inertia cubic nonlinearity featured by electro-optical crystals in the THz range, Photonics, 7, pp. 1–7.
26. Sajadi M., Wolf M. & Kampfrath T. (2017). Transient birefringence of liquids induced by terahertz electric-field torque on permanent molecular dipoles, Nature Communications, 8, pp. 1–9.
27. Bodrov S., Sergeev Y., Murzanev A. & Stepanov A. (2017). Terahertz induced optical birefringence in polar and nonpolar liquids, Journal of Chemical Physics, 147, pp. 084507.
28. Tcypkin A. *et al.* (2021). Giant third-order nonlinear response of liquids at terahertz frequencies, Physical Review Applied, 15, pp. 054009.
29. Tcypkin A. *et al.* (2019). High Kerr nonlinearity of water in THz spectral range, Optics Express, 27, pp. 10419.
30. Francis K. J. G., Chong M. L. P., Zhang X.-C. & E Y. (2020). Terahertz nonlinear index extraction via full-phase analysis, Optics Letters, 45(20), pp. 5628–5631.
31. Boyd R. W. (2008). *Nonlinear Optics*. Academic Press.

32. Artser I. *et al.* (2022). Radiation shift from triple to quadruple frequency caused by the interaction of terahertz pulses with a nonlinear Kerr medium, Scientific Reports, 12, pp. 1–8.
33. Novelli F., Hoberg C., Adams E., Klopf J. & Havenith M. (2022). Terahertz pump–probe of liquid water at 12.3 THz, Physical Chemistry Chemical Physics, 24, pp. 653–665.
34. Novelli F. *et al.* (2020). Nonlinear terahertz transmission by liquid water at 1 THz, Applied Sciences, 10, pp. 5290.
35. Sheik-Bahae M., Said A. A., Wei T. H., Hagan D. J. & Van Stryland E. W. (1990). Sensitive measurement of optical nonlinearities using a single beam, IEEE Journal of Quantum Electronics, 26, pp. 760–769.
36. Zheng X. *et al.* (2015). Characterization of nonlinear properties of black phosphorus nanoplatelets with femtosecond pulsed Z-scan measurements, Optics Letters, 40, pp. 3480.
37. Cao H., Linke R. A. & Nahata A. (2004). Broadband generation of terahertz radiation in a waveguide, Optics Letters, 29, pp. 1751.
38. Yeh K. Lo, Hebling J., Hoffmann M. C. & Nelson K. A. (2008). Generation of high average power 1 kHz shaped THz pulses via optical rectification, Optics Communications, 281, pp. 3567–3570.
39. Melnik M., Vorontsova I., Putilin S., Tcypkin A. & Kozlov S. (2019). Methodical inaccuracy of the Z-scan method for few-cycle terahertz pulses, Scientific Reports, 9, pp. 1–8.
40. Drozdov A. A., Kozlov S. A., Sukhorukov A. A. & Kivshar Y. S. (2012). Self-phase modulation and frequency generation with few-cycle optical pulses in nonlinear dispersive media, Physical Review A, 86, pp. 053822.
41. Vicario C. *et al.* (2015). Intense, carrier frequency and bandwidth tunable quasi single-cycle pulses from an organic emitter covering the Terahertz frequency gap, Scientific Reports, 5, pp. 1–8.
42. Kislin D. A., Knyazev M. A., Shpolyanskii Y. A. & Kozlov S. A. (2018). Self-action of nonparaxial few-cycle optical waves in dielectric media, JETP Letters, 107, pp. 753–760.
43. Kozlov S. A. & Samartsev V. V. (2013). *Fundamentals of Femtosecond Optics.* Elsevier.
44. Tapia A. K. G., Yamamoto N., Ponseca C. & Tominaga K. (2011). Charge carrier dynamics of ZnSe by optical-pump terahertz-probe

spectroscopy, 36th International Conference on Infrared, Millimeter, and Terahertz Waves, pp. 1–2.

45. Tcypkin A. *et al.* (2018). Experimental estimate of the nonlinear refractive index of crystalline ZnSe in the terahertz spectral range, Bulletin of the Russian Academy of Sciences: Physics, 82, pp. 1547–1549.

46. Kozlov S. A. *et al.* (2018). Self-focusing does not occur for few-cycle pulses, Journal of Physics: Conference Series, 1092, pp. 012066.

47. Tcypkin A. *et al.* (2019). High Kerr nonlinearity of water in THz spectral range, Optics Express, 27, pp. 10419.

48. Yang K. H., Richards P. L. & Shen, Y. R. (1971). Generation of far-infrared radiation by picosecond light pulses in LiNbO3, Applied Physics Letters, 19, pp. 320–323.

49. Kuznetsov S. A. *et al.* (2010). Development and characterization of quasi-optical mesh filters and metastructures for subterahertz and terahertz applications, Key Engineering Materials, 437, pp. 276–280.

50. Watanabe A., Saito H., Ishida Y., Nakamoto M. & Yajima T. (1989). A new nozzle producing ultrathin liquid sheets for femtosecond pulse dye lasers, Optics Communications, 71, pp. 301–304.

51. Tcypkin A. *et al.* (2021). Giant third-order nonlinear response of liquids at terahertz frequencies, Physical Review Applied, 15, pp. 054009.

52. Ezerskaya A. A., Ivanov D. V., Kozlov S. A. & Kivshar Y. S. (2012). Spectral approach in the analysis of pulsed terahertz radiation, Journal of Infrared, Millimeter, and Terahertz Waves, 33, pp. 926–942.

53. Ho P. P. & Alfano R. R. (1979). Optical Kerr effect in liquids, Physical Review A, 20, pp. 2170–2187.

54. Markowicz P. P. *et al.* (2004). Modified Z-scan techniques for investigations of nonlinear chiroptical effects, Optics Express, 12, pp. 5209.

55. Weber M. J. (2018). *Handbook of Optical Materials*. CRC Press.

56. Azarenkov A. N., Al'tshuler G. B., Belashenkov N. R. & Kozlov S. A. (1993). Fast nonlinearity of the refractive index of solid-state dielectric active media, Quantum Electronics, 23, pp. 633–655.

57. Kittel C. & Holcomb D. F. (1967). Introduction to solid state physics, American Journal of Physics, 35(6), pp. 547–548.

58. Guselnikov M. S., Zhukova M. O. & Kozlov S. A. (2022). Inertia of the oscillatory mechanism of the giant nonlinearity of optical

materials in the terahertz spectral range, Journal of Optical Technology, 89, pp. 3–12.

59. Kozlov S. A. (1995). The classical theory of dispersion of high-intensity light, Optics and Spectroscopy, 4106, pp. 404–409.

60. Hale G. M. & Querry M. R. (1973). Optical constants of water in the 200-nm to 200-μm wavelength region, Applied Optics, 12(3), pp. 555–563.

61. Thrane L., Jacobsen R. H., Uhd Jepsen P. & Keiding S. R. (1995). THz reflection spectroscopy of liquid water, Chemical Physics Letters, 240, pp. 330–333.

62. Schatzberg P. (1967). On the molecular diameter of water from solubility and diffusion measurements, Journal of Physical Chemistry, 71, pp. 4569–4570.

63. Kell G. S. (1967). Precise representation of volume properties of water at one atmosphere, Journal of Chemical & Engineering Data, 12, pp. 66–69.

64. Ghalgaoui A. *et al.* (2020). Field-induced tunneling ionization and terahertz-driven electron dynamics in liquid water, Journal of Physical Chemistry Letters, 11, pp. 7717–7722.

65. Ambrose D., Sprake C. H. S. & Townsend R. (1975). Thermodynamic properties of organic oxygen compounds XXXVII. Vapour pressures of methanol, ethanol, pentan-1-ol, and octan-1-ol from the normal boiling temperature to the critical temperature, The Journal of Chemical Thermodynamics, 7, pp. 185–190.

66. Huang S. *et al.* (2009). Improved sample characterization in terahertz reflection imaging and spectroscopy, Optics Express, 17(5), pp. 3848–3854.

67. Moutzouris K. *et al.* (2013). Refractive, dispersive and thermo-optic properties of twelve organic solvents in the visible and near-infrared, Applied Physics B, 116(3), pp. 617–622.

68. van der Bruggen B., Schaep J., Wilms D. & Vandecasteele C. (1999). Influence of molecular size, polarity and charge on the retention of organic molecules by nanofiltration, Journal of Membrane Science, 156, pp. 29–41.

69. Kuntman A. & Baysal B. M. (1993). Estimation of thermodynamic parameters for poly(ethyl methacrylate)/isopropyl alcohol system from intrinsic viscosity measurements, Polymer, 34, pp. 3723–3726.

70. Fileti E. E., Castro M. A. & Canuto S. (2008). Calculations of vibrational frequencies, Raman activities and degrees of depolarization for complexes involving water, methanol and ethanol, Chemical Physics Letters, 452, pp. 54–58.
71. Wilmink G. J. *et al.* (2011). Development of a compact terahertz time-domain spectrometer for the measurement of the optical properties of biological tissues, Journal of Biomedical Optics, 16, pp. 047006.
72. Yomogida Y., Sato Y., Nozaki R., Mishina T. & Nakahara J. (2010). Comparative dielectric study of monohydric alcohols with terahertz time-domain spectroscopy, Journal of Molecular Structure, 981, pp. 173–178.
73. Rheims J., Köser J. & Wriedt T. (1997). Refractive-index measurements in the near-IR using an Abbe refractometer, Measurement Science and Technology, 8, pp. 601.
74. Buntrock R. E. (2013). Review of *The Merck Index: An Encyclopedia of Chemicals, Drugs, and Biologicals*, 15th edition, Journal of Chemical Education, 90, pp. 1115.
75. Sun T. F., ten Seldam C. A., Kortbeek P. J., Trappeniers N. J. & Biswas S. N. (2006). Acoustic and thermodynamic properties of ethanol from 273.15 to 333.15 K and up to 280 MPa, Physics and Chemistry of Liquids, 18, pp. 107–116.
76. Upshur M. A. *et al.* (2016). Vibrational mode assignment of α-pinene by isotope editing: one down, seventy-one to go, Journal of Physical Chemistry A, 120, pp. 2684–2690.
77. George D. K. & Markelz A. G. (2012). Terahertz Spectroscopy of Liquids and Biomolecules, in *Terahertz Spectroscopy and Imaging*, Springer, pp. 229–250.
78. Woldegeorgis A. *et al.* (2018). THz induced nonlinear effects in materials at intensities above 26 GW/cm2, Journal of Infrared, Millimeter, and Terahertz Waves, 39, pp. 667–680.
79. Abraham E., Freysz E., Degert J. & Cornet M. (2014). Terahertz Kerr effect in gallium phosphide crystal, Journal of the Optical Society of America B: Optical Physics, 31(7), pp. 1648–1652.
80. Korpa C. L., Tóth G. & Hebling J. (2016). Interplay of diffraction and nonlinear effects in the propagation of ultrashort pulses, Journal of Physics B: Atomic, Molecular and Optical Physics, 49, pp. 035401.

81. Boyd R. (2008). Ultrafast and intense-field nonlinear optics, Nonlinear Optics, pp. 561–587.
82. Arnaud N. & Georges J. (2001). Investigation of the thermal lens effect in water–ethanol mixtures: composition dependence of the refractive index gradient, the enhancement factor and the Soret effect, Spectrochimica Acta Part A: Molecular and Biomolecular Spectroscopy, 57, pp. 1295–1301.
83. Ge X. & Lu D. (2017). Molecular polarizability of water from local dielectric response theory, Physical Review B, 96, pp. 075114.
84. Wilkes Z. W. et al. (2009). Direct measurements of the nonlinear index of refraction of water at 815 and 407 nm using single-shot supercontinuum spectral interferometry, Applied Physics Letters, 94, pp. 211102.

© 2024 World Scientific Publishing Company
https://doi.org/10.1142/9789811265648_0006

Chapter 6

Spatio-temporal Control of Intense Laser-Induced Terahertz/X-ray Emission from Water

Hsin-hui Huang,[1,2] Saulius Juodkazis,[2,3,4] Takeshi Nagashima,[5]
Koji Hatanaka,[1,2,6,7,8]

[1]*Research Center for Applied Sciences, Academia Sinica, Taiwan*
[2]*Optical Sciences Center and ARC Training Centre in Surface Engineering
for Advanced Materials (SEAM), School of Science, Computing and
Engineering Technologies, Optical Sciences Center, Swinburne
University of Technology, Australia*
[3]*Tokyo Tech World Research Hub Initiative (WRHI), School of Materials
and Chemical Technology, Tokyo Institute of Technology, Japan*
[4]*Institute of Advanced Sciences, Yokohama National University, Japan*
[5]*Faculty of Science and Engineering, Setsunan University, Japan*
[6]*College of Engineering, Chang Gung University, Taiwan*
[7]*Department of Materials Science and Engineering,
National Dong-Hwa University, Taiwan*
[8]*Center for Optical Research and Education (CORE),
Utsunomiya University, Japan*

6.1 Introduction: Intense Laser Interaction with Water

Ultrashort sub-100 fs laser pulses at moderate focusing to focal spots of tens of micrometers reach the average intensity at PW/cm^2 with pulse energy of <1 mJ. Such conditions provide very promising and practical light-matter interaction. Tunnelling ionization becomes dominant over the avalanche of multiphoton electron generation. Transfer of matter to plasma state is established, which takes place within the timescale of the pulse duration. Different states of plasma can be generated at such conditions, where the under-dense plasma is still transparent for laser pulses which followed and focused in gases (or rarefied condensed solids/liquids) along the propagation of the laser pulse at focal volume. A point-like energy deposition is realised when laser pulses irradiate liquid or solid targets, causing almost instantaneous ionization. The deposited (absorbed) energy inside the interaction region, hence, has the energy density at J/m^3 with the pressure at N/m^2 driving surface-localized explosion through laser ablation on the surface. This energetic event is considered to be the source of X-ray (keV), visible and infrared (IR, eV), and terahertz (THz) (meV) emission (Fig. 6.1). Such an on-demand delivery source of wide-range photon radiation using liquid targets, which can be easily manipulated by having specific additives for specific emission bands, is one target of our research, which is described in the following sections. Detailed discussion of the

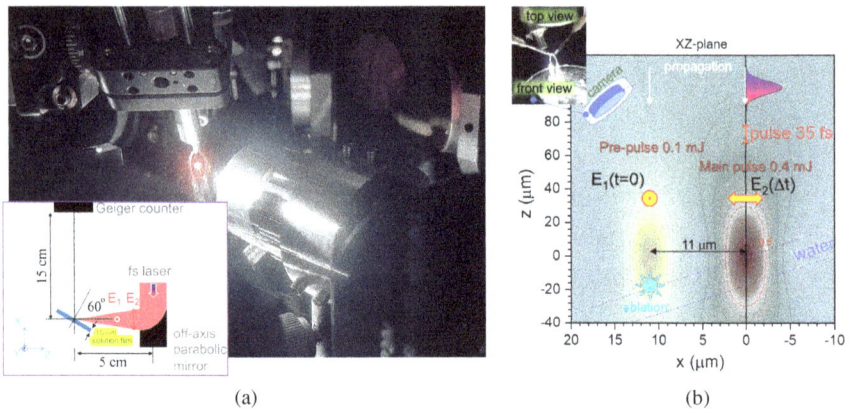

(a)   (b)

▲ Fig. 6.1. (a) Photo of the actual setup used for the dual X-ray and THz source under two-pulse irradiation. Inset shows schematics for the X-ray detection modality. (b) Geometry of interaction region. Optical Gaussian intensity distributions for the pre-pulse (energy E_1) and the main pulse (E_2) are shown for the actual focusing conditions. Inset shows a micro-thin water flow. Separation of pre- and main pulse by 11 μm was required for the most intense THz generation.

experimental data are published elsewhere [1–6]. In this chapter, basic concepts are described with supporting data sets and focus on geometrical aspects of light-matter interaction.

6.2 THz and X-ray Simultaneous Emission

6.2.1 Experimental

The experimental setup is as shown in Fig. 6.2(a). A flat solution flow of milli-Q water with thickness smaller than 20 μm was prepared using a nozzle. The flow rate was set at < 70 mL/min and regulated by a circulation pump. The nozzle was mounted on a rotational and 3D-automatic stage

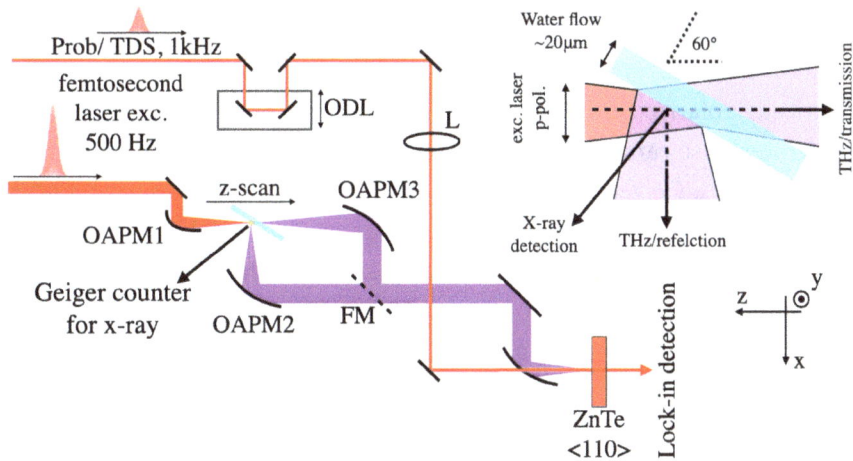

▲ Fig. 6.2. The experimental setup for the simultaneous measurements of THz (time-domain spectroscopy, TDS) and X-ray (Geiger counter). The femtosecond laser pulses (>35 fs, 800 nm, 500 Hz, horizontally polarized to the solution flow surface, i.e., p-pol.) were focused onto the solution flow. ODL - optical delay line for TDS. L - plano-convex lens (f = 50 cm). The thickness of the <110>ZnTe crystal for TDS was 1 mm. Parent focal lengths for OAPMs (off-axis parabolic mirrors) were f = 50.8 mm (OAPM1, 1-inch diameter), 101.6 mm (OAPM2, 2-inch), 152.4 mm (OAPM3, 2-inch), and 101.6 mm (OAPM4, 2-inch with a hole in its center for the probe). The distance between the laser focus and Geiger counter was 12 cm. FM - flip-folding mirror for TDS measurements in transmission direction. The inset shows the close-up of the solution surface, the laser-water interaction region. Reprinted from Ref. [4] with permission.

and its position was finely controlled by a home-made LabView code [7]. Transform-limited femtosecond laser pulses (λ = 800 nm, > 35 fs, 1 kHz, linearly polarized) were separated as two beams with different polarizations by a half-wave plate and a polarization beam splitter. Horizontally and vertically polarized pulses were defined as the excitation pulse for X-ray/ THz generation and the probe pulse for the time-domain spectroscopy (TDS) for THz measurements [8, 9], respectively. The repetition rate of the excitation pulses was modulated by a wheel chopper for TDS measurements, and the pulses were tightly focused in air onto the solution flow surface by using an off-axis parabolic mirror (OAPM, 1-inch diameter, effective focus length f = 50.8 mm, reflection angle of 90 degrees, and numerical aperture NA = 0.25). The incident angle of the excitation pulses along the z-axis to the solution normal was fixed at 60 degrees for the highest X-ray emission as reported previously [1]. Under these experimental conditions, each excitation pulse at 500 Hz repetition rate irradiates the fresh and flat solution flow. Experiments on double-pulse excitation with a pre-pulse (vertically polarized, 0.1 mJ/pulse, 4.6 ns in advance of the main excitation pulse) were also carried out using an optical delay line. X-ray intensity was measured by a Geiger counter with different size of apertures in air under atmospheric pressure (1 atm) at room temperature (296 K). Its observation angle was 15 degrees to the solution normal toward the excitation side and its distance from the laser focus was 12 cm. Therefore, it is certain that the Geiger counter detects only X-ray, not α- nor β-ray. THz signals were collected on the reflection (30 degrees to the solution normal, 90 degrees to the laser incident direction) and the transmission (along the excitation z-axis) directions. As conventional TDS, THz emission and the probe pulses after variable optical delay were focused to a 1 mm-thick ⟨110⟩-oriented ZnTe crystal by an OAPM and a plano-convex lens, respectively.

6.2.2 Z-scan Experiments

One experiment for THz emission enhancements was performed under double-pulse excitation. Fig. 6.3 shows simultaneous emission of X-ray (a) and THz (b,c) with the pre-pulse (0.1 mJ/pulse, vertically polarized, s-pol.)

X-ray

THz reflection

(a)

(b)

THz transmission

(c)

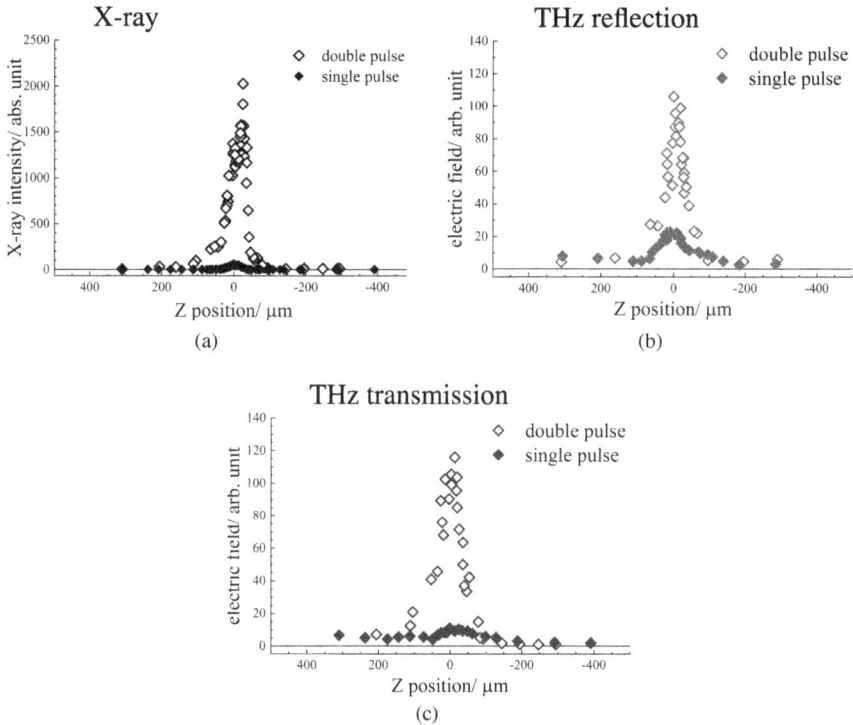

▲ Fig. 6.3. Z-position-dependent intensities of (a) X-ray, (b) THz emission in the reflection and (c) THz transmission under single-pulse (solid diamond) and double-pulse (open diamond) excitation conditions. For double-pulse excitation, the laser intensities for the main excitation pulse (horizontally polarized, p-pol.) and the pre-pulse (vertically polarized, s-pol., 4.6 ns in advance of the main pulse) were 0.4 mJ/pulse and 0.1 mJ/pulse, respectively. Reprinted from Ref. [4] with permission.

with the delay time at 4.6 ns in advance to the main pulse (0.4 mJ/pulse, horizontally polarized, p-pol.) irradiation. X-ray emission apparently showed an intensity enhancement under double-pulse excitation, as expected [2, 3]. One additional peak at the downstream side was also clearly discernible, as reported [7, 10]. With a time delay of 4.6 ns between the pulses, the initial processes of water film ablation and transient surface roughening under action of capillary forces and micro-droplet formation (mist) at the position close to the initial location of the solution surface, all induced by the pre-pulse irradiation, caused a more effective coupling of

the main pulse with such a modified solution surface (laser ablation). The enhancement was caused by multiple scattering, local refocusing of light by droplets and the perturbed surface that resulted in the X-ray intensity enhancement, which can reach an order of magnitude greater and is useful for practical applications. THz emission in the reflection (Fig. 6.3(b)) was also enhanced about five times, and the profile width under double-pulse excitation became narrower at 52 μm as compared with the 106 μm width under single-pulse excitation. THz emission profile along the z-axis under double-pulse excitation showed only a single peak at the center as in the case of single-pulse excitation, which was different from the profile of X-ray emission with the additional peak at the downstream side. This is consistent with the requirement of thermal gradients lasting ~1 ps for ~1 THz emission, which are less likely on a fragmented water film, while X-ray emission is maintained by hot plasma and geometrical factors are less important. In the case of THz emission in the transmission, the profile showed an apparent change (Fig. 6.3(c)); namely, the intensity was enhanced more than ten times, the profile along the z-axis changed to a single peak from the profiles with a dip at its center, and transient surface roughness, droplet (mist) formation and hole formation on the solution flow were expected at a delay time of 4.6 ns [2]. These initial processes of laser ablation induced by the pre-pulse irradiation may cause enhancement of THz emission especially in the transmission direction.

6.3 Spatio-temporal Control Based on Laser Ablation

Under double pulse irradiation conditions with the pre-pulse, several phenomena such as transient surface roughness (shock wave generated), droplet (mist) and ablation-pit formation on the surface of solution flow are considered as laser ablation induced by the pre-pulse irradiation. Intensity of the pre-pulse is high enough to induce laser ablation but lower than that of the main pulse. Related dynamics of such surface morphological changes have been well characterized by time-resolved surface scattering imaging techniques with spatial and temporal resolutions at a few tens of nanometers and a few tens of picoseconds, respectively [11]. With this knowledge, THz emission can be well manipulated on demand by utilizing the spatial and temporal offsets for the pre-pulse irradiation to the main pulse irradiation. Spatio-temporal effects of two-pulse irradiation for THz emission are described below in detail.

6.3.1 Experimental

The pre-pulse spatial offset setup is as shown in Fig. 6.4(a). The spatial offsets along x- and y-axes, Δx and Δy, are applied to the pre-pulse irradiation in the experiments. One steering mirror for the pre-pulse alignment is automatically controlled by a piezo-transducer system to change the pre-pulse focus position with micron-order precision. The origin, (Δx, Δy) = (0, 0), is defined experimentally when the pre-pulse and the main pulse overlap co-linearly and with a Newton ring-like pattern in far field to the transmission side as an interference pattern with the pre-pulse and main pulse focuses in air. The focus position of the pre-pulse with offsets is experimentally confirmed by the imaging setup described as follows.

Imaging experiments for the laser focus from the side (along x-axis) and the top (6° tilted from y-axis) are as shown in Fig. 6.4(b), and carried out in two different methods with an objective lens (10×) and CMOS cameras with filters for IR cut and intensity control in the visible region. One is with the

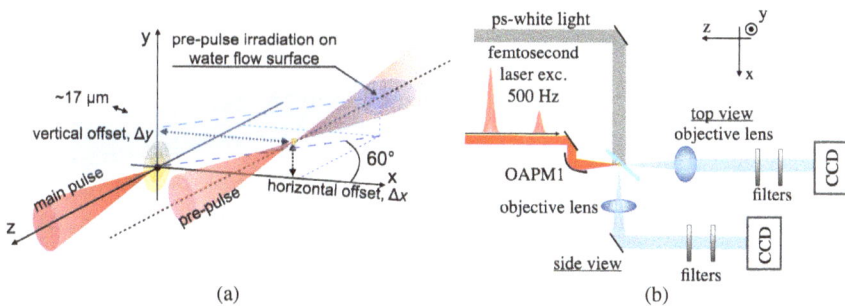

▲ Fig. 6.4. (a) Schematic diagram for the horizontal and vertical offsets, Δx and Δy, for the pre-pulse irradiation. (b) Schematic diagram of time-resolved shadowgraphy under the pre-pulse irradiation with picosecond white light back illumination and time-integrated luminescence images under single- (only with the main pulse) and double-pulse excitation condition. Reprinted from Ref. [6] with permission.

pre-pulse and white light continuum (~1 ps, 580 ± 30 nm, as a strobe light) converted from the main pulse with a water cell. With this method, transient refractive index changes and/or scattering due to pre-plasma formation and/or laser ablation induced by the pre-pulse irradiation can be visualized. Another imaging is with the pre-pulse and the main pulse, which visualizes the interaction of the main pulse with the water flow with structures prepared by the pre-pulse irradiation. The exposure time for the camera setting was fixed at 2 ms for single-shot imaging. In this mode of image acquisition, all the emission in broadband spectra by the two-pulse irradiation of the water flow is time-integrated. In addition to the imaging, time-integrated UV-visible emission spectroscopy is also performed with a commercially available spectrometer, which is position dependent with a fiber input.

The above described detailed geometry of experiments highlights the importance of direct imaging and characterization of the interaction volume. At the used focusing corresponding to NA ≈ 0.125, there was no filamentation of fs pulse, and air ionization (dielectric breakdown) with plasma formation was contained within depth-of-focus or double Rayleigh length. Fig. 6.5 shows time-integrated plasma emission in air under different irradiation conditions and time evolution of the shockwave front for single pre-pulse. Shocked air volume and its temporal evolution were clearly captured using shadowgraphy. In water (Fig. 6.5(c)), there was shock front in air as well as in water. Exact determination of laser pulse interaction on the front side of water flow was clearly determined [6].

Nonlinear character of light-matter interaction at the used intensities of 1–50 PW/cm^2 is shown in Fig. 6.6. Transmittance of the single main pulse was different in air and water by an order of magnitude (Fig. 6.6(a)). The two-photon absorption (TPA) coefficient β [cm/PW] was estimated from departure of transmission from the linear law. If TPA is the main absorption mechanism, a simple expression for the transmittance is applicable $\ln(1 + \beta L \times I_p)/(\beta L)$ and is plotted as a fit in Fig. 6.6(b). For the length of focal region in air $L \approx 60$ μm, one finds $\beta = 2.85$ cm/PW, and in the pre-ionized air by pre-pulse $E_1 = 0.2$ mJ at the 4.7 ns delay before main pulse, $\beta = 19.55$ cm/PW for $l = 800$ nm. The time delay of 4.7 ns was set for

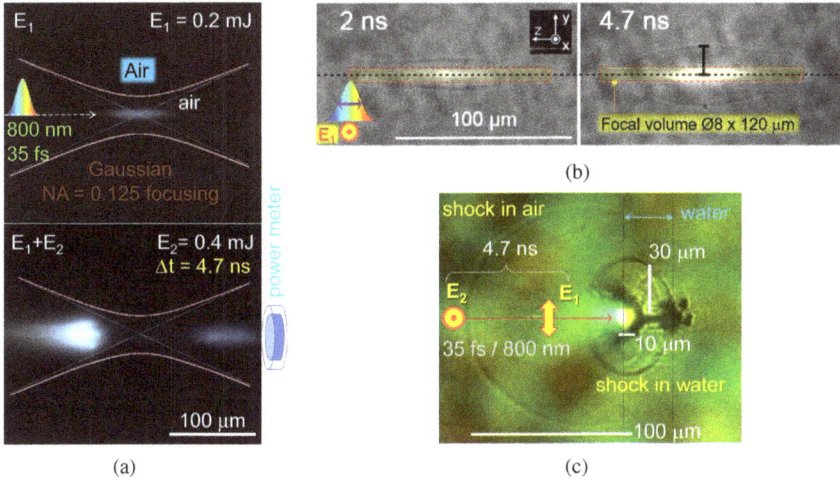

(a)

(b)

(c)

▲ Fig. 6.5. Visualization of interaction volume in air and in front of water flow. (a) Time-integrated photoluminescence from the Rayleigh length $2z_R = \pi r^2/\lambda$ for one- and two-pulse irradiation along the same optical axis (pre- and main pulse are aligned); focal spot radius $r = 0.61\,\lambda/NA$. (b) Shadowgraphy of single-pulse action in air and (c) two-pulse irradiation of water. Conditions: $E_1 = 0.1$ mJ and $E_2 = 0.4$ mJ for water irradiation in (c). Reprinted from Ref. [6] with permission.

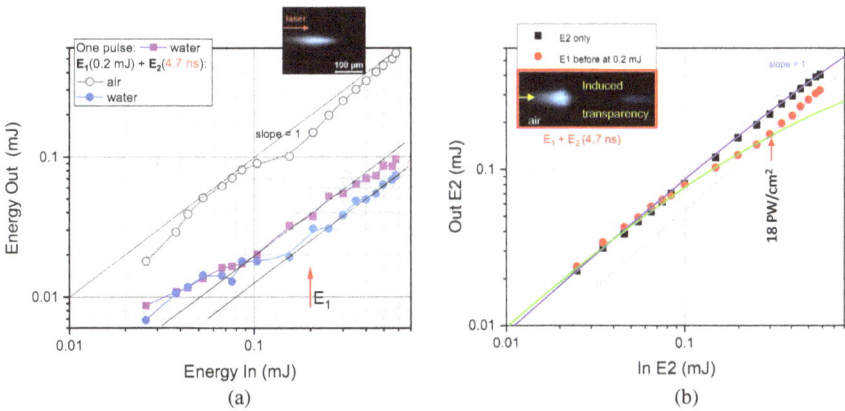

(a)

(b)

▲ Fig. 6.6. Visualization of nonlinear light-matter interaction. (a) Transmittance of the main pulse E_2 in air and water departs from linear dependence (slope = 1) at increasing pulse energy. (b) The two-photon absorption coefficient β is obtained from fit $\ln(1 + \beta L \times I_p)/(\beta L)$, where L is the interaction length (the depth of focus in air), and I_p is pulse intensity (for $E_p = 0.3$ mJ, $I_p = 18$ PW/cm^2). Inset shows time-integrated photoluminescence image for two-pulse irradiation (see Fig. 6.5(a)). Scales of optical images are the same in (a) and (b).

stronger THz emission from air/water targets and was found empirically (this time separation is also dependent on spatial offsets of the pulses).

6.3.2 THz Control by the Pre-Pulse Irradiation with Spatio/Temporal Offsets

With this setup, intensities of X-ray and THz were measured simultaneously with spatial offsets for the pre-pulse irradiation along the x- and y-axes, while the temporal delay for the main pulse irradiation was at 4.6 ns (Fig. 6.7). In the case of X-ray, a resultant component of the pre-pulse-induced plasma (under-dense) at the center position may enhance the X-ray emission dramatically, such that the X-ray emission intensity under double-pulse irradiation shows its concentric nature. On the other hand, in the case of THz emission, its highest emission intensity was clearly observed when the spatial offsets for the pre-pulse irradiation was at $(\Delta x, \Delta y) = $ (11 μm, 0 μm). This asymmetric nature of THz emission can be attributed to laser ablation by the pre-pulse irradiation to the tilted water flow.

Under double-pulse excitation condition with the spatio-temporal offset for the pre-pulse at $(\Delta x, \Delta y, \Delta t) = $ (11 μm, 0 μm, 4.7 ns), its representative result is as shown in Fig. 6.8(a). As the delay time between the two pulses

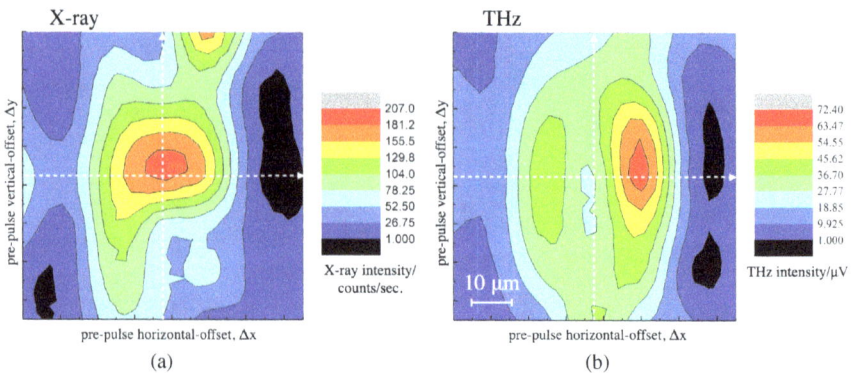

▲ Fig. 6.7.   Intensity dependence on the pre-pulse spatial offset along x- and y-axes when the delay time for the main pulse is Δt = 4.6 ns. (a) X-ray and (b) THz in the transmission side. The main pulse always irradiates the center position in these plots.

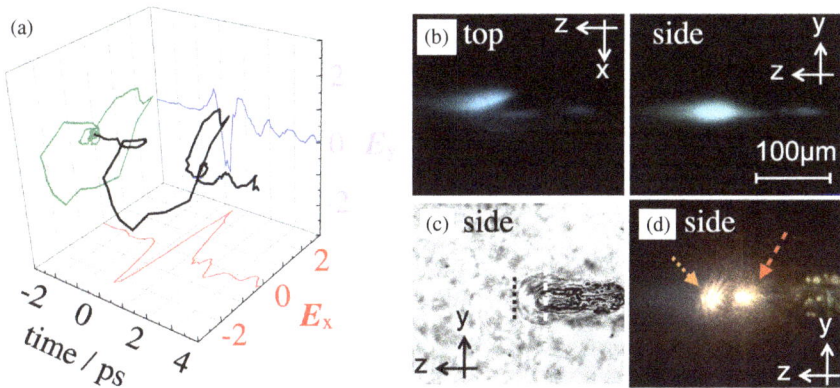

▲ Fig. 6.8. (a) The TDS signal waveform in the transmission side measured at $\Delta t = 4.7$ ns with the pre-pulse (y-pol.) at 0.2 mJ per pulse and the main pulse (x-pol.) at 0.4 mJ per pulse. The pre-pulse spatial offset is fixed at $(\Delta x, \Delta y) = (11\ \mu m, 0\ \mu m)$. (b) Time-integrated luminescence images (top and side) at $\Delta t = 4.7$ ns in air (without the water flow). (c) Time-resolved shadowgraphy with the water flow from the side with picosecond white light back illumination. The dotted lines in black and white represent the shockwave front and the flow surface, respectively. (d) Time-integrated luminescence image at $\Delta t = 4.7$ ns with the water flow. The arrows in orange (dotted line) and red (dashed line) in the luminescence image indicate the light emission at the shockwave front and the laser focus on the flow surface, respectively. Reprinted from Ref. [6] with permission.

increases, the polarization status of THz emission dynamically changes to circular polarization (left-handed circular, LHC, clockwise when it is observed from the later time to the earlier time) as shown, from horizontally linear under single-pulse excitation [6]. Furthermore, with the spatio-temporal offset, THz emission intensity increases 1,500 times higher than in single-pulse irradiation as $|E|^2$-intensity in discrete Fourier transform spectra over the 0–3 THz region [5]. Considering the report on the energy conversion efficiency from 800 nm to THz wave with a ZnTe(110) crystal at 1.25×10^{-5} [12], the photon number-based conversion efficiency under double-pulse irradiation can be estimated to be 7.1×10^{-3}.

Fig. 6.8(b) shows time-integrated luminescence images at $\Delta t = 4.7$ ns only with air (without the water flow). The pre-pulse spatial offset is fixed at $(\Delta x, \Delta y) = (11\ \mu m, 0\ \mu m)$. In the case only with air, though a stretched rugby ball-like blue emission is observed at the focus under single-pulse

excitation, the spatial distribution of the blue emission, which reflects the intensity spatial distribution of the incident main pulse, shows distorted shapes under double-pulse irradiation with the pre-pulse with the spatial offset. As the delay time increases, the spatial distribution changes [6]. This is due to shockwave expansion from the volume of air ionization due to the pre-pulse irradiation. The expansion velocity of the shockwave front is estimated to be about 7 km/s initially at $\Delta t = 2.0$ ns and slows down to about 1 km/s at $\Delta t = 14.7$ ns [6]. With the water flow as shown in the time-resolved shadowgraphy in Fig. 6.8(c) and the luminescence images in Fig. 6.8(d), another shockwave formation due to laser ablation by the pre-pulse excitation to the water flow (therefore this shockwave front carries ablation plume of ionized/atomized water), its expansion in front of the flow, and formation of cavitation bubbles in the flow are clearly observed.

Experiments were extended to reveal the effects of spatial offsets between the two pulses for the control of THz emission. Fig. 6.9 shows the results of TDS measurements for the transmission side. The delay time between the two pulses is fixed at 4.7 ns. As the horizontal offset becomes large along the x-axis from the condition in Fig. 6.8(a) at $(\Delta x, \Delta y) = (11$ μm, 0 μm), THz polarization changes from elliptical to horizontally linear. As the vertical offset coincides with the fixed horizontal offset at $\Delta x = 11$ μm, THz polarization changes from horizontally linear back to circular but with opposite handedness, LHC (in blue) to RHC (in red). Neither an apparent THz intensity enhancement nor THz circular polarization were observed when the pre-pulse irradiates the water flow with the horizontal offset with negative values. On the other hand, in the case only with air without the water flow, the tendency of the THz polarization dependent on the pre-pulse spatial offset is homogeneous in its radial direction (as expected from its symmetry) [6]. This indicates that the asymmetric characteristic of the tilted water flow is important.

The symmetry of X-ray generation was apparently broken for THz emission under two-pulse exposure. Another manifestation of the broken symmetry was observation of a circularly polarized THz emission when excitation was linearly polarized. This was qualitatively explained

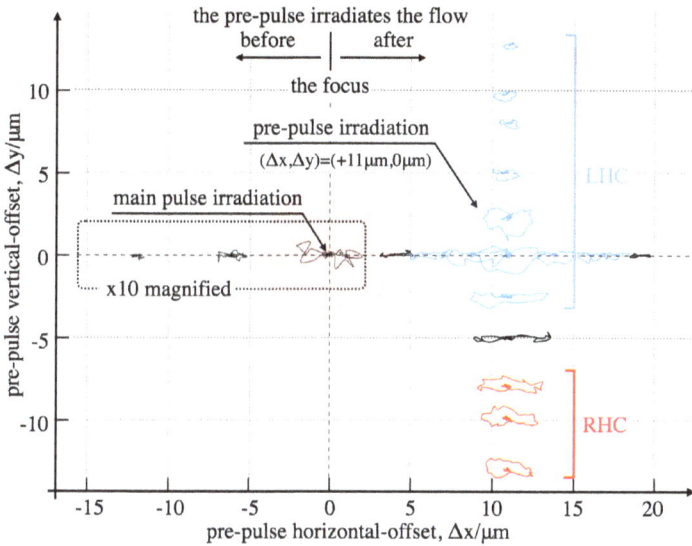

▲ Fig. 6.9. Projections of TDS signal waveforms to x-y plane (as the plot on the green panel in Fig. 6.8(a)) under double-pulse excitation with delay time at 4.7 ns on the transmission direction. The plots in blue and red represent left-handed circular (LHC) and right-handed circular (RHC), respectively. The plots in black represent almost linearly polarized emission. The red and blue circles represent the effective area (8 μm in diameter at the focus) of the main and the pre-pulse irradiation, respectively. Reprinted from Ref. [6] with permission.

by interaction of mass density gradient generated by the pre-pulse and coupled with the wake field velocity of electrons. Since the wake field is not propagating and has fixed geometry (generated by the main pulse), the position of the pre-pulse becomes important since it defines material transfer from the ablation site toward the main pulse on the water flow surface by a shock wave as well as by ablation plume. The mass density pushed toward the trajectory and irradiation point of the main pulse is transferred into the electron density (charge) gradient by ionization ∇n_e and is coupled with the electron velocity v_e in the wake field velocity. Their cross product $\nabla n_e \times v_e$ defines the handedness of circular current and, hence, polarization of emitted THz. Depending on which part of the standing wave wake field the mass (charge) is injected from prepulse, the polarization of the current can be RHC or LHC. Duration of charge transient defines spectral content of the emission: 1 THz emission corresponds to the 1 ps lifetime of rotational current. THz emission measured in transmission and reflection had different handedness, which is consistent with phase change upon reflection from the surface of water flow. This is also consistent with direct imaging of the two-pulse irradiation and pins the light-matter interaction to the (pre-)surface area of the water flow.

6.4 Summary

Detailed imaging of two-pulse irradiation of water flow by shadowgraphy and photoluminescence clearly revealed interaction to be located within the focal volume of the used optics. The axial extent of the focal region was ~ 100 μm long and ~ 10μm in diameter at the focus, which was placed on the surface of water flow. X-ray and THz emission can be optimized using two-pulse irradiation. For THz, spatial separation plays an important role for the most intense THz emission at the optimized separation between two pulses (few ns), while for X-ray, temporal separation was the only useful control parameter.

The conversion efficiency of photon numbers from the 800 nm laser to 1 THz under double-pulse excitation condition was calculated to be 7.1 × 10^{-3} [6] as shown in Table 6.1, where the conditions for the estimate are as follows; target sample was milli-Q water with thickness

▼ Table 6.1.   Conversion efficiency based on photon numbers from near-IR (800 nm) laser to 1 THz.

	Single Pulse	Double Pulse*
Air	1.6 × 10^{-6}	1.6 × 10$^{-5†}$
Water	4.7 × 10^{-6}	7.1 × 10^{-3}

* The intensities for the main and the pre-pulse are 0.4 mJ/pulse and 0.2 mJ/pulse, respectively.
† in air, with the pre-pulse offset at (Δx, Δy) = (0, +33 μm) [13].
(As a reference, the conversion efficiency with a $\langle 110 \rangle$ ZnTe crystal is 4.7 × 10^{-2} [12].)

of 17±3 μm or with ambient air at room temperature and 1 atm. The femtosecond laser pulses have the shortest pulse width of 35 fs at 800 nm. The main pulse was at 0.4 mJ/pulse and the pre-pulse was at 0.2 mJ/pulse. The delay time between the main and the repulse was $\Delta t = 4.7$ ns. The conversion efficiency of photon numbers from the 800 nm laser to X-ray at 3–20 keV under double-pulse excitation condition was calculated to be 1.3×10^{-8} [2], where the conditions for the estimate are slightly different from the THz case aforementioned. The target sample was cesium chloride aqueous solution with concentration of 4 mol/dm³ and flow thickness of 40 μm. The femtosecond laser pulses are 260 fs, negatively chirped at 780 nm. Main pulse was at 0.3 mJ/pulse and the pre-pulse was at 0.06 mJ/pulse. The delay time between the main and the repulse was $\Delta t = 4$ns [14].

Acknowledgments

The authors thank Profs. Eugene G. Gamaly and Vladimir Tikhonchuk for extensive discussions of the light-matter interactions. S. J. is grateful for partial support by the ARC DP190103284, LP190100505 and JST CREST JPMJCR19I3 grants. T. N. is grateful for the support by JSPS KAKENHI Grant Number 20K05371. K. H. is grateful for the support by the National Science and Technology Council (NSTC, formerly known as MOST) of Taiwan (107-2112-M-001-014-MY3, 110-2112-M-001-054), the Cooperative Research Program of "Network Joint Research Center for Materials and Devices", Nanotechnology Platform (Hokkaido University), and the Collaborative Research Projects of the Laboratory for Materials and Structures, Institute of Innovative Research (Tokyo Institute of Technology). K. H. also acknowledges the Japan Science and Technology Agency (JST) PRESTO (Precursory Research for Embryonic Science and Technology) Program (SAKIGAKE, Innovative use of light and materials/ life) for its support on the original project on X-ray/THz wave simultaneous emission, "Ultra-wide band light conversion by controlling structures of microdroplets and ultrashort laser pulse (2009–2013)" and for the laser facilities for the current project.

References

1. Hatanaka K., Ida T., Ono H., Matsushima S., Fukumura H., Juodkazis S. & Misawa H. (2008). Chirp effect in hard X-ray generation from liquid target when irradiated by femtosecond pulses, Optics Express, 16(17), pp. 12650–12657.
2. Hatanaka K., Ono H. & Fukumura H. (2008). X-ray pulse emission from cesium chloride aqueous solutions when irradiated by double-pulsed femtosecond laser pulses, Applied Physics Letters, 93(6), pp. 064103.
3. Hatanaka K. & Fukumura H. (2012). X-ray emission from CsCl aqueous solutions when irradiated by intense femtosecond laser pulses and its application to time-resolved XAFS measurement of I⁻ in aqueous solution, X-Ray Spectrometry, 41(4), pp. 195–200.
4. Huang H.-H., Nagashima T., Hsu W.-H., Juodkazis S. & Hatanaka K. (2018). Dual THz wave and X-ray generation from a water film under femtosecond laser excitation, Nanomaterials, 8(7), pp. 523.
5. Huang H.-H., Nagashima T., Yonezawa T., Matsuo Y., Ng S. H., Juodkazis S. & Hatanaka K. (2020). Giant enhancement of THz wave emission under double-pulse excitation of thin water flow, Applied Sciences, 10(6), pp. 2031.
6. Huang H.-H., Juodkazis S., Gamaly E. G., Nagashima T., Yonezawa T. & Hatanaka K. (2022). Spatio-temporal control of THz emission, Communications Physics, 5(1), pp. 134.
7. Hsu W.-H., Masim F. C. P., Balčytis A., Juodkazis S. & Hatanaka K. (2017). Dynamic position shifts of X-ray emission from a water film induced by a pair of time-delayed femtosecond laser pulses, Optical Express, 25(20), pp. 24109–24118.
8. Wu Q. & Zhang X.-C. (1995). Free-space electro-optic sampling of terahertz beams, Applied Physics Letters, 67, pp. 3523.
9. Lee Y.-S. (2009). *Principles of Terahertz Science and Technology.* Springer.
10. Huang H.-H., Nagashima T., Hsu W.-H., Juodkazis S. & Hatanaka K. (2018). Dual THz wave and X-ray generation from a water film under femtosecond laser excitation, Nanomaterials, 8(7), pp. 523.

11. Hatanaka K., Itoh T., Asahi T., Ichinose N., Kawanishi S., Sasuga T., Fukumura H. & Masuhara H. (1998). Time-resolved surface scattering imaging of organic liquids under femtosecond KrF laser pulse excitation, Applied Physics Letters, 73(24), pp. 3498–3500.
12. Blanchard F., Razzari L., Bandulet H.-C., Sharma G., Morandotti R., Kieffer J.-C., Ozaki T., Reid M., Tiedje H. F., Haugen H. K. & Hegmann F. A. (2007). Generation of 1.5 µJ single-cycle terahertz pulses by optical rectification from a large aperture ZnTe crystal, Optics Express, 15(20), pp. 13212–13220.
13. Huang H.-H., Nagashima T. & Hatanaka K. (2023). Shockwave-based THz emission in air, Optical Express, 31(4), pp. 5650–5661.
14. Hatanaka K., Ono H. & Fukumura H. (2008). X-ray pulse emission from cesium chloride aqueous solutions when irradiated by double-pulsed femtosecond laser pulses, Applied Physics Letters, 93(6), pp. 064103.

© 2024 World Scientific Publishing Company
https://doi.org/10.1142/9789811265648_0007

Chapter 7

Terahertz Wave Generation from Water Lines

Jianming Dai, Yuxuan Chen, Yuhang He

Center for Terahertz Waves and School of Precision Instrument and Opto-electronic Engineering, Tianjin University, China

7.1 Introduction

In recent years, liquids such as water [1, 2], liquid nitrogen [3], liquid metal [4], ethanol, and acetone [5, 6] have been experimentally proven to be promising broadband terahertz (THz) sources due to their higher molecular density and lower ionization threshold compared to those in gases. Benefiting from the fluidity of liquids, liquid flows of different geometries have been utilized as laser targets to provide a fresh zone between adjacent pump laser pulses. There are mainly two geometries of liquid flow proposed previously: plane geometry (such as liquid films [2] and liquid flat jets [5–9]) and cylindrical geometry (e.g., liquid lines [10–14]). A common problem that occurred with the plane geometry is the relatively low efficiency of THz wave coupling out of liquid media. Particularly, the THz waves are strongly absorbed by the flat film of water in the lateral directions. Besides, the total internal reflection of the THz wave at the flat water-air interface reduces the coupling efficiency of THz signals significantly. To improve the efficiency of THz wave coupling out of water flows, water lines with cylindrical geometry are proposed as liquid-based THz sources.

Two factors, the total internal reflection as well as the absorption inside the water line, both of which depend on the geometry of the water flow, may play important roles in the coupling of the emitted THz wave. The efficiencies of THz wave coupling out of liquid media and THz wave propagation transfer functions from emitter to detector are different between planar and cylindrical flows. However, the mechanism of THz wave generation may be consistent in different liquid flows [15], which means that the THz waves generated from water lines and water films share many characteristics. In this chapter, we will introduce the characteristics of the THz wave generated from water lines, including its dependences on the diameter of the water line, laser pulse width, relative position between the water line and laser beam axis, detection angle, and pump laser energy.

7.2 Experimental Setup

In practical terms, water lines with various diameters in a large range (Fig. 7.1(a)) can be easily obtained using commercially available dispensing nozzles with different inner diameters (Fig. 7.1(b)), while water films with different thicknesses are not easily attained due to the limitation in the fixed dimension of jet nozzles. Furthermore, stable and smooth surfaces of water lines are seemingly easier to be formed in comparison with those of water films. A schematic of the water circulation system for generation of water lines is shown in Fig. 7.1(c). First, water from reservoir 1 (R1) is pumped into reservoir 2 (R2) by a peristaltic pump. Then the pressure in R2 squeezes water into the needle, forming a water line. When the pressure in R2 is stable, the flow rate of the water line is determined by the speed of the peristaltic pump. Thus, the flow rate of the water line can be set at different values and kept stable in the experiment.

There are two feasible ways to reduce the impact of the pre-pulse in water. First, using a laser of lower repetition rate (for example, 10–20 Hz) is an effective method to avoid the influence of the pre-pulse. Another way is to increase the flow rate of the water line to make each laser pulse (for a 1-kHz repetition rate, the pulse-pulse separation is 1 ms) interact with

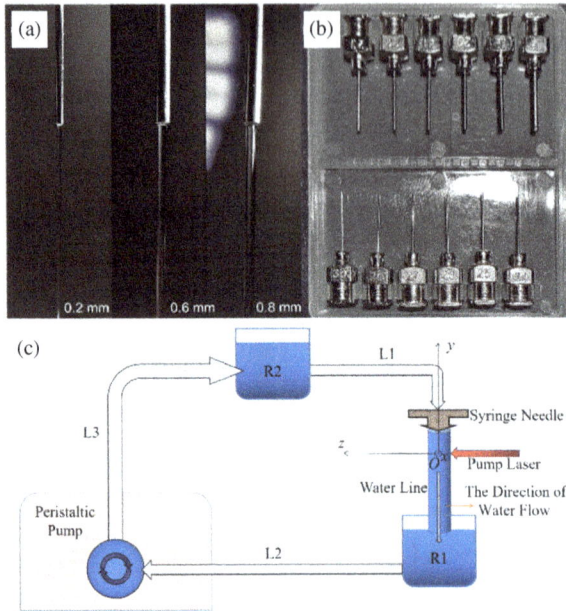

▲ Fig. 7.1. (a) Water lines with diameters of 0.2, 0.6, and 0.8 mm. (b) Industrial dispensing nozzles with different inner diameters. (c) Schematic of the water circulation system used to generate stable water lines.

a fully refreshed volume in the water line. In order to avoid the impact of adjacent pump laser pulses, the flow rate of the water line is usually kept at 6~7 mm/ms, ensuring each optical pump pulse ionizes a completely refreshed water volume. Many of the previous works on THz wave generation in liquids also used laser pulses with a 1-kHz repetition rate, since the sampling rate is relatively higher than those with 10–20 Hz laser systems. It is noteworthy that the water temperature can affect the efficiency of THz wave generation [6]. Three major approaches are utilized to minimize the impact of laser heating on the THz wave generation efficiency: first, a relatively large flow rate of the water line (6~7 mm/ms) is needed to avoid the local heating effect; second, the total volume of the water (the total heat capacity) in the circulation system needs to be large enough to ensure that the overall heating effect is negligible; and third, a chiller is used to control the water temperature directly at the expense of system complexity [6].

To detect the THz wave generated from water lines, a THz time-domain spectroscopic system can be used, as shown in Fig. 7.2(a). A commercial femtosecond Ti:sapphire amplified laser system with 800-nm central wavelength and 1-kHz repetition rate is commonly used in experiments. Different laser pulse durations can be achieved by changing the pre-chirp of the output laser pulse. Experimentally, negatively chirped pulses can be used to maximize THz wave generation efficiency [2]. The laser beam is split into pump and probe beams. The pump beam is focused into a water line by a two-inch effective focal length (EFL) lens. The main consideration for the choice of focal length is the plasma length. To eliminate the possibility of THz wave generation from air plasma, the laser-induced plasma needs to be fully included inside the water line, which means the Rayleigh range needs to be much smaller than the diameter of the water line. Therefore, a lens with a relatively high numerical aperture is employed to focus the laser beam into the water line, ensuring that the laser-induced plasma is within the water line. The estimated double Rayleigh range of the laser beam at the focal point is about 28 µm when the focal length is 50.8 mm, satisfying the above requirement. In experiments, if the focal length is too short, the arrangement of the experimental setup would be more difficult. In addition, the divergence angle of the THz wave coupled out of the water line will be too large for the parabolic mirror behind to collect a sufficient amount of THz radiation into the detector. On the other hand, the tighter focusing gives rise to better overall generation efficiency at lower pump pulse energy but would make THz wave generation from the plasma saturate earlier as pump pulse energy increases. Hence, a two-inch EFL lens (or parabolic mirror) is commonly used to focus the laser beam into the water line. The THz wave generated from the water line is collected by an off-axis parabolic mirror (PM1) with six-inch EFL. A high-resistivity silicon wafer (Si) is used as a long-pass filter to block the residual pump light. The THz wave passes through the silicon wafer and is refocused by another off-axis parabolic mirror (PM2) with four-inch EFL. An ITO-coated glass plate is used to combine the optical probe beam and THz wave. A 1-mm thick, <110> cut ZnTe crystal followed by a balanced detector (BD) is used to detect the THz waveform. It is noteworthy that a larger collecting angle

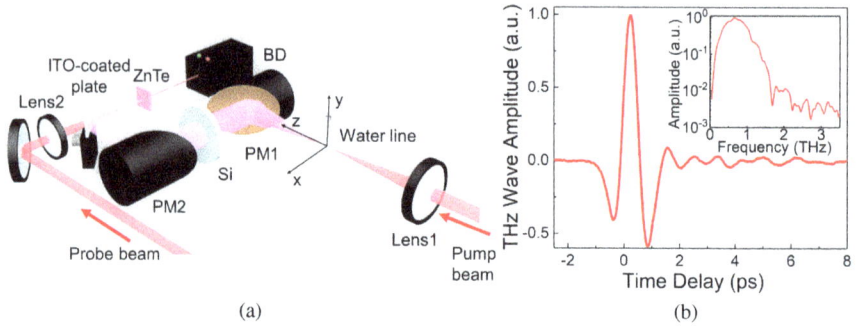

▲ Fig. 7.2.   (a) The experimental setup. (b) THz waveform generated from a water line. Inset: The Fourier transform spectrum of the THz wave. Adapted from Ref. [13] with permission.

can be obtained by using a short EFL parabolic mirror (PM1), which may increase the collecting efficiency of THz radiation. However, the intention for using a parabolic mirror with long EFL for the collection of the THz wave at the emitter is to focus the THz beam at the detection crystal to a smaller beam spot size to enhance the detection efficiency (or say, to get better signal-to-noise ratio). In the experimental setup, the combination of the six-inch EFL and the four-inch parabolic mirrors forms a telescope, and changing the ratio between the two focal lengths can effectively tune the THz beam size at the detector. Considering the frequency-dependent beam size of the generated THz wave, a larger THz beam focal spot at the ZnTe crystal will result in the loss of some frequency components. Furthermore, THz emission from a relatively short plasma has a very large angular distribution. In fact, independent experimental results show that THz emission from water lines can also be collected and detected from the sideway of the plasma (with a parabolic mirror to collect THz radiation in the direction perpendicular to the pump beam axis) using electrooptic (EO) detection [14, 16], implying that THz emission from water lines in the current case cannot be completely collected by any off-axis parabolic mirrors. In the experimental setup shown in Fig. 7.2(a), only the emitted THz wave in the forward direction is collected. However, by rotating the detection system (or by changing the propagating direction of the pump beam) around the water line, THz radiation can be detected from any direction. We define that the laser pulse propagates along the z-direction and the water line flows along the y-direction. The detection angle represents the angle between the z-axis and the optical axis of PM1. The nozzle is mounted on a two-dimensional translation stage so that the position of the water line can be precisely controlled. The typical waveform of the THz wave collected in the forward direction from a 0.2-mm water line is shown in Fig. 7.2(b), and its corresponding Fourier transform spectrum is shown in the inset.

7.3 Dependence of THz Emission Efficiency on Pump Pulse Duration and Diameter of Water Line

Since water lines with different diameters can be easily achieved using nozzles with different inner diameters, systemic investigation on the optimal water line diameter for THz wave generation from liquid water is made possible. In this section, we will introduce the impact of pump pulse duration and diameter of the water line on THz wave generation.

First, like the water film scheme, the THz radiation depends on the pump laser pulse durations [2]. To explore the dependence of THz radiation on the optical pulse duration, different laser pulse durations are achieved by changing the pre-chirp of the laser pulse (in this experiment we use negatively chirped laser pulses) while the pump laser pulse energy is kept at 0.4 mJ, and measured by an SHG intensity autocorrelator. The THz amplitude from a 0.2-mm water line is measured and plotted in Fig. 7.3(a) as the pulse duration increases from 98 fs to 590 fs. The THz amplitude increases at first and then decreases with the pulse duration. The optimal pulse duration for THz wave generation from the 0.2-mm water line is ~400 fs. The trend of the THz amplitude variation with pulse duration agrees with the results in Fig. 7.4(a), which is ascribed to the cascade ionization in laser-irradiated liquid water [11]. In the theory presented in Ref. [11], tunnel and multiphoton ionization processes provide seed electrons for the cascade ionization process. Besides, it takes a fixed time for each cascade

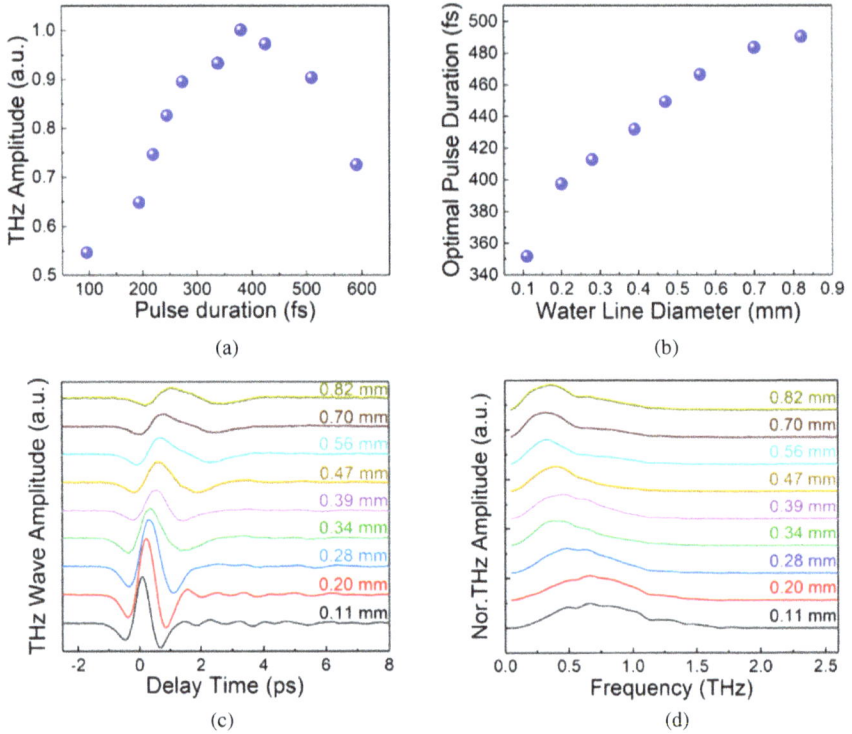

▲ Fig. 7.3. (a) THz amplitude as a function of pump pulse duration for 0.20 mm water line. (b) Optimal pulse duration for water lines with different diameters. (c) THz waveforms generated from water lines with different diameters. (d) Corresponding Fourier transform spectra. Reprinted from Ref. [13] with permission.

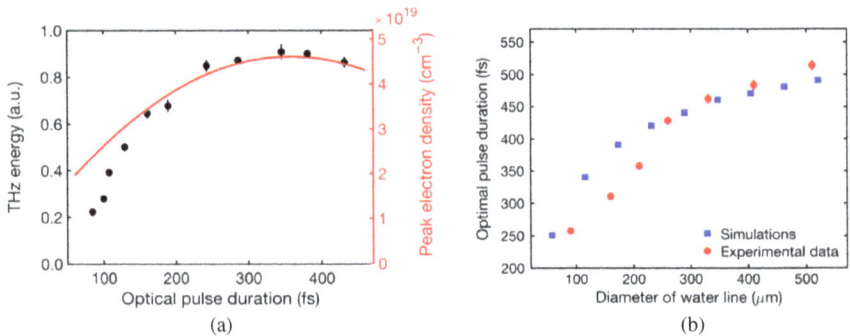

▲ Fig. 7.4. (a) Dependences of THz energy and peak electron density on optical pulse duration for a 210-μm water line. (b) Optimal optical pulse duration vs the diameter of the water line. Reprinted from Ref. [11] with permission.

process to finish. Thus, from one seed electron, the cascade process can produce electrons no more than $2\tau_p/\tau_i$ in number. A longer pulse duration enables more cascade processes to take place, leading to a higher electron density. However, the laser intensity is inversely proportional to the pulse duration when the pulse energy is fixed. As tunnel ionization highly depends on laser intensity and provides most of the seed electrons for the cascade process, the electron density in plasma starts to decrease when the pulse duration is too long. The trade-off between tunnel and cascade ionization processes leads to the optimal pulse duration for the highest electron density in water, as shown by the red solid line in Fig. 7.4(a). Since higher electron density in water leads to stronger THz radiation, the optimal pulse duration for THz radiation generation coincides with that for electron generation.

Benefiting from the availability of water lines with different diameters, we can systematically investigate the optimal parameters for THz wave generation. A set of nozzles with different inner diameters are used to produce water lines with different diameters. It is noteworthy that the optimal pulse duration is not the same for different diameters of water lines. In this experiment, we also negatively chirp initial laser pulses to achieve different pulse durations and measure the optimal pulse duration for each water line using an SHG intensity autocorrelator. We plot the optimal pulse duration versus the diameter of the water line in Fig. 7.3(b). The optimal pulse duration increases with the diameter of the water line, and the variation tends to be flattened for larger diameters of water lines. The result in Fig. 7.3(b) agrees with the simulation and experimental results shown in Fig. 7.4(b) [11].

Under the optimal pulse durations for THz wave generation in water lines with different diameters, the THz waveforms and their corresponding spectra are different. Fig. 7.3(c) shows THz waveforms generated from water lines with different diameters. For each water line, the positions of laser focus and laser pulse duration are optimized individually to achieve the maximal THz peak amplitude. THz wave generated from a water line with 0.20-mm diameter has the highest peak amplitude and it decreases as the diameter deviates from 0.20 mm. Furthermore, we can observe an

obvious THz pulse peak timing delay and pulse width broadening when the water line diameter increases. Such a pulse peak timing delay is attributed to the expansion of the optical path length as the water line diameter increases. Fig. 7.3(d) shows the corresponding Fourier transform spectra. A frequency redshift occurs when the water line diameter increases, which corresponds to the THz pulse broadening. There are two reasons for this effect: First, the generated THz waves have longer optical path lengths in thicker water lines. Considering the absorption coefficient of water increases with frequency in the range of 0.1–10 THz [17], high-frequency components of the generated THz wave experience higher absorption in the thicker water lines. The absence of high-frequency components leads to the spectrum redshift and THz pulse broadening. Second, the optimized pump pulse durations for thicker water lines are longer, leading to a longer oscillation period of the ponderomotive force-induced dipole due to the slower varying intensity profile of laser pulse, which may directly broaden the generated THz pulse.

7.4 Dependence of THz Radiation on Relative Position Between Water Line and Laser Beam

It has been experimentally and theoretically demonstrated that, for the water film scheme, the optimal incident angle of the laser beam is ~65° for the forward THz wave emission [7]. Furthermore, THz signals can be hardly detected from any detection angle when the laser beam is normally incident into the water film. The results can be attributed to the orientation of the dipole, whose moment is oriented along the laser propagation direction. By rotating the water film, the dipole moment is changed since the incident angle of the laser beam is varied, leading to the tuning of the efficiency of THz wave coupling out of the water film. However, even at the optimal incident angle, most of the THz energy cannot be coupled out of the water film due to total internal reflection at the water-air interface and the absorption of water [7]. In contrast, the water line scheme can significantly reduce the total internal reflection owing to its cylindrical water-air interface, and enables the investigation into the sideway emission of THz radiation [14, 16]. Thus, in the case of the water line scheme, the full angular distribution of THz radiation from laser-induced plasma in liquid water can be experimentally revealed. The incident angle of the laser beam is determined by the relative position between the water line and the laser beam axis along the x-direction. The optimal relative positions for the maximum collecting efficiency of the THz wave generated from the water line are different for forward and lateral emission schemes. In this section, we will introduce the dependence of THz radiation on the relative position between the water line and the laser beam for forward emission scheme, while that for the lateral emission scheme will be discussed in the next section.

For the forward emission scheme, the dependence of THz radiation on the relative position between the water line and the laser beam has already been demonstrated with water lines with different diameters. Fig. 7.5(a) shows the peak THz wave amplitude as the 200-μm water line is scanned across the laser beam axis along the x-direction [10]. When the laser axis is at the center of the water line, virtually no THz pulse can be collected. As the center of the water line deviates from the laser axis, the THz wave amplitude first increases and then decreases with positive and negative maxima located around $x = -/+(60\sim70$ μm$)$, respectively. When the laser focus is outside the water line boundary but not far away from the water line surface ($x > 100$ μm), part of the laser beam still hits the water line and the ionization in water still occurs. Thus, the THz signal from water is also recorded even though the laser focus is not within the water line completely. As the water line keeps moving away from the laser focus, the ionization in water and thereby the THz radiation is gradually weakened and eventually vanishes, which explains the falling trend when the focus is outside the water line boundary. The polarity of the THz amplitude is inverted as the position of the water line on the x-direction varies from negative to positive. In particular, as

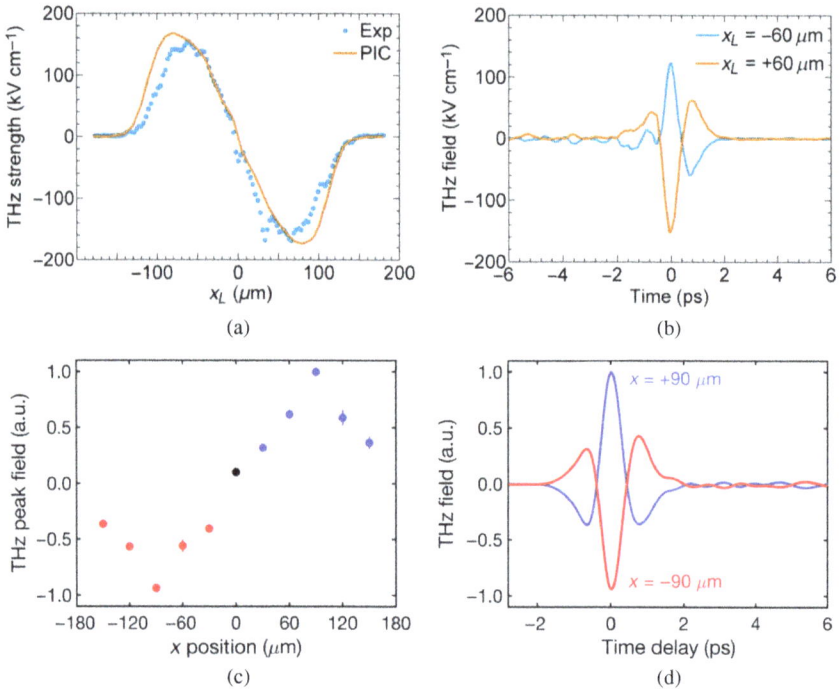

▲ Fig. 7.5. (a), (c) THz peak amplitudes as a function of x positions for 200-μm and 260-μm water lines, respectively. Reprinted from Ref. [10] with permission. (b), (d) THz waveforms at $x = \pm60$ μm and $x = \pm90$ μm for 200-μm and 260-μm water lines, respectively. Reprinted from Ref. [11] with permission.

the center of the water line oppositely deviates from the laser axis by the same distance, the waveforms of the THz radiation are reversed while the absolute values of THz amplitude are nearly identical. Fig. 7.5(b) shows two THz waveforms with inversed polarity in comparison to each other as the laser axis deviates from the center of the water line by +60 μm and −60 μm. Similar results are obtained when a 260-μm water line is used, as shown in Figs. 7.5(c) and (d) [11]. In the experiment, a weak THz signal is detected at $x = 0$ μm, which is represented by the black dot in Fig. 7.5(c). This coincides with the case where a water film is pumped by a laser beam at normal incidence. When the water line is shifted away from the zero position in the x-direction the THz signal increases, which presents the same trend as that shown in Fig. 7.5(a). The THz electric field is maximized at $x = \pm 90$ μm, but the THz waveforms have opposite polarities, as shown in Fig. 7.5(d). The results shown in Figs. 7.5(c) and (d) can be explained by the dipole model. Moving the water line along the x-direction is essentially equivalent to the change of incident angle of the laser beam by rotating the water film [7]. Mirrored x-positions of the water line lead to the opposite dipole orientation in water, which is similar to rotating the water film with opposite angles, resulting in inverted THz waveforms. When the laser axis is at the center of the water line, the dipole in water is oriented along the direction of laser propagation, emitting weak THz radiation in the forward direction. However, the shift of the water line along the x-direction results in a stronger THz signal in the forward direction because the dipole moment is tilted.

The aforementioned experimental results show that the maximal THz amplitude occurs at $x = \pm 60 \sim 70$ μm for the 200-μm water line and at $x = \pm 90$ μm for the 260-μm water line, respectively. For water lines with other diameters, the dependence of THz radiation on the water line position remains the same [12]. Fig. 7.6(a) shows the detected THz peak amplitude as the water line moves along the x-axis. In the experiment, the pulse duration is kept consistent for water lines with different diameters. As the water lines with different diameters move along the x-axis, the THz amplitudes follow the same trend as that shown in Figs. 7.5(a) and (c). For water lines with diameters of 0.2 mm, 0.3 mm, 0.4 mm, and 0.5 mm, the peak values of the THz signal appear at $x = \pm 0.07$ mm, ± 0.11 mm, ± 0.14 mm, and ± 0.16 mm, respectively. The THz time-domain waveforms at the optimal x-positions for water lines with different diameters are shown in Figs. 7.6(b)–(e). For each water line at the positive and negative optimal x-positions, the THz waveforms are inverted. However, as the water line diameter increases from 0.2 mm to 0.5 mm, the generated THz amplitude decreases, which agrees with the results in Fig. 7.3(c).

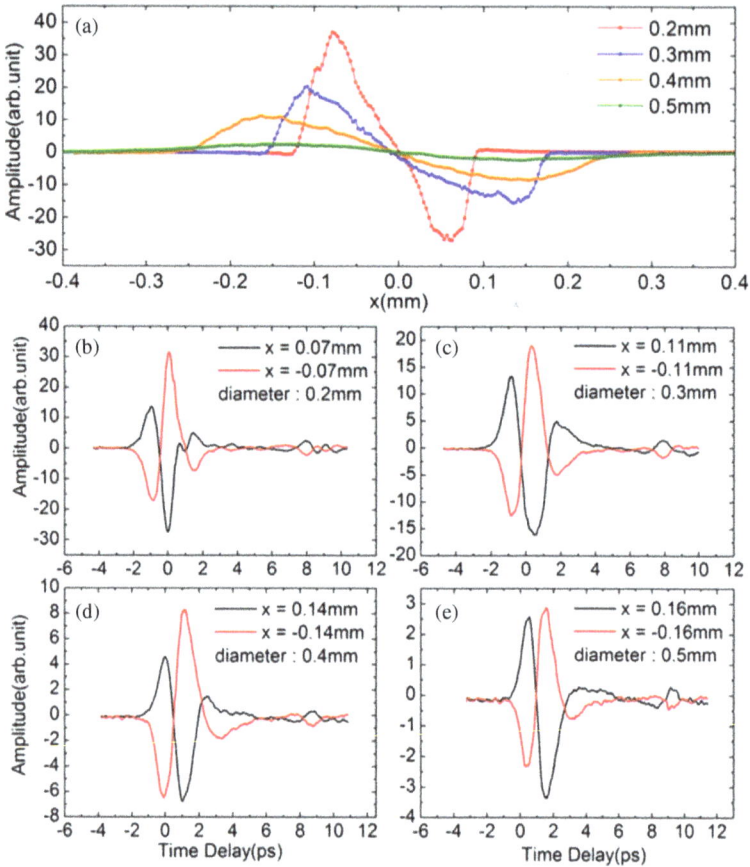

▲ Fig. 7.6. (a) THz peak amplitude as a function of x-position with the water line diameter of 0.2 mm, 0.3 mm, 0.4 mm, and 0.5 mm, respectively. (b)–(e) THz waveforms at the optimal x-positions for water lines with different diameter. Reprinted from Ref. [12] with permission.

The results on the optimal water line positions for water lines with different diameters can be fitted linearly, as shown in Fig. 7.7. An empirical rule concluded from the linear fitting is that the peak values of the THz amplitude will appear at the position $x = \pm0.68 \times R$, where R is the radius of the water line. In other words, during the optimization of the THz signal, the optimal deviation for the highest THz wave coupling efficiency is about

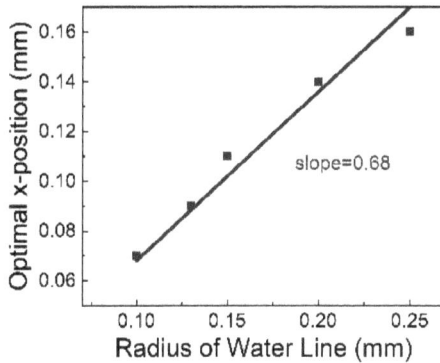

slope=0.68

▲ Fig. 7.7.   Optimal water line position as a function of the radius of the water line. Solid line: linear fitting of the experimental results.

68% of the radius of the water line, independent of the actual diameter of the water line. In this case, the angle of the dipole orientation with respect to the z-axis is calculated to be ~12° as the water line is located at the optimal position for any diameter, which is relatively small in comparison to that in the water film at optimal incident angles. Considering that the THz wave mainly emits in directions perpendicular to the dipole moment, a dipole with an orientation angle closer to 90° can contribute a stronger THz wave in the forward direction. It is noteworthy that, as the pump beam deviates from the center of the water line, the optical intensity at the focus is diminished due to additional astigmatism arising from the cylindrical water line surface. As a result, the overall THz wave generation efficiency might be slightly reduced, even though the detected signal increased. Therefore, we cannot obtain a stronger THz signal by simply enlarging the deviation of the pump beam from the center of the water line, although the angle of the dipole orientation increases with it. As the orientation angle of the dipole is only ~12°, the strongest THz amplitude cannot be obtained from the forward direction, even though the position of the water line is optimized for detection along the forward direction [10, 12]. The angular distribution of the THz radiation from the water line will be discussed in Section 7.6.

7.5 Lateral THz Wave Emission from Water Lines

It has been proved that the lateral detection scheme is more efficient than the forward scheme [14, 16]. It is necessary to introduce the impact of the relative position between the water line and the laser beam on the THz wave collection efficiency for the lateral scheme. Since THz emission from liquid water is very sensitive to the relative position between the water line and the laser beam axis, an experiment about the impact of relative position on THz emission is performed to fully characterize the lateral THz wave emission [14]. The 110-μm water line is scanned across the laser beam axis along the x-axis. The dependence of THz energy on the x-direction displacement of the water line is measured at the detection angle of 90° with respect to the laser axis. As shown in Fig. 7.8(a), the dependence exhibits one peak on each side of the origin. For convenience, we define the THz signal peak at $x < 0$ as signal A and that at $x > 0$ as signal B, with their peak positions as position A ($x = -30$ μm) and position B ($x = +50$ μm), respectively. The trend of two peaks is seemingly counterintuitive. The THz signal generated from the area $x > 0$ was supposed to be lower than that from the other side, since it has a longer optical path in water leading to stronger water absorption, as shown in Fig. 7.8(b). However, our experimental result shows that signal B is even greater than signal A. In order to give a clearer overview in the THz signal change, a set of consecutive THz waveforms are taken at the detection angle of 90° as the water line moves along x-direction from -100 μm to 80 μm step by step with 10-μm step size, as shown in Fig. 7.8(c). Here, typical Fourier transform spectra of selected THz waveforms are shown in Fig. 7.8(d), which only include the data between -60 μm and 60 μm to avoid the influence of THz wave emission from air plasma. Different from the forward scheme, the polarity of the THz wave does not change in sign and the absolute values of the THz amplitude are not identical for water line positions symmetric about the origin. As the water line moves from the x < 0 region toward the $x > 0$ region, the overall pulse width is broader while the THz amplitude becomes larger. Overall, the bandwidths of the Fourier transform spectra in the $x < 0$ region are broader than those in the $x > 0$

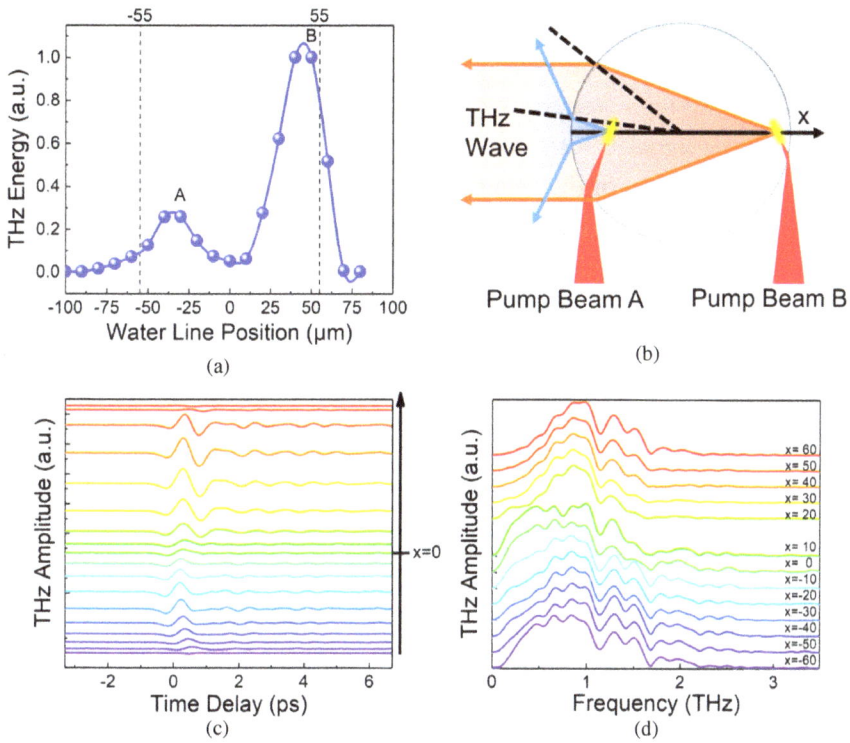

▲ Fig. 7.8. (a) The dependence of THz energy on the water line position along the x-direction measured at 90°. Vertical dashed lines indicate the boundary of the water line. (b) Schematic of THz wave propagation in the water line. (c) Variation of THz waveform as a function of wave line position. (d) The Fourier transform spectra at different water line positions in the x-direction. Reprinted from Ref. [14] with permission.

region. The mechanism of the counterintuitive "two peaks" THz energy variation tendency and the waveform variation can be attributed to the trade-off between THz absorption in water and THz collecting efficiency.

In the THz frequency range, liquid water has a large frequency-dependent absorption coefficient. The absorption coefficient increases with frequency in the region of 0.1–3.0 THz [17], implying that a longer optical path in water leads to a narrower bandwidth of the THz spectrum. A comparison of the THz waveforms and their spectra of signal A and signal B are shown in Figs. 7.8(c) and (d), respectively. The observed

spectrum narrowing as the water line moves toward the right direction can be explained by the frequency-dependent absorption coefficient. However, the loss of the high-frequency components and the increase of the peak amplitude of signal B seem to conflict with each other. Nevertheless, the collection efficiency of the generated THz wave coupling out of the liquid medium cannot be neglected. The collection efficiency is highly dependent on the divergence angle of the THz wave due to the limited collection angle of the parabolic mirror (about 19°). A simplified model based on the THz wave refraction on the water-air surface is utilized to describe the variation of divergence angle as a function of the x-direction displacement of the water line, which is sketched in Fig. 7.8(b). In the $x < 0$ region, the cylindrical water surface can be treated as a negative cylindrical lens. The generated THz wave will be diverged by the cylindrical water surface, leading to a low collection efficiency. The divergence angle decreases as the water line moves toward the origin from the $x < 0$ region, which increases the collection efficiency. On the other hand, as the focal point moves toward the origin, the increase of the optical path in water leads to higher absorption loss of the THz wave. The trade-off between absorption loss and collection efficiency of the THz wave coupling out of the water line determines the optimal water line position for THz collection in the $x < 0$ region, corresponding to the water line position of signal A. On the contrary, the water line surface can be seen as a positive cylindrical hyper-hemisphere lens in the $x > 0$ region. The generated THz wave will be converged or collimated by the hyper-hemisphere surface. The divergence angle decreases as the water line moves in the positive direction. Likewise, trade-off between the absorption in water and the collection efficiency determines the optimal water line position for the highest THz signal in the $x > 0$ region, corresponding to the water line position of signal B. Moreover, the refractive index of water in the THz band is also a function of frequency. The refraction index decreases with frequency in the 0.1–3.0 THz region but remains above 2 in the 0.1–1.5 THz band [17]. The divergence angles of different frequency components directly depend on the refraction index of THz wave in liquid water. For low-frequency components the divergence angles are smaller than those for high-frequency components in the $x > 0$ region and vice versa in the $x < 0$ region, due to the abnormal dispersion of water in the THz band. Considering the collecting angle is relatively small, only the THz wave with a divergence angle within ±9.5° can be collected by the parabolic mirror. In this case, the collection efficiencies of different frequency components are different at a fixed water line position. In other words, the optimal water line positions of different frequency components for their highest collection efficiencies are different. For example, Fig. 7.8(d) shows a blueshift of the THz spectrum when the water line position increases from +20 μm to +50 μm. This phenomenon is related to the abovementioned abnormal dispersion in the 0.1–3.0 THz region. The refraction indices of low-frequency components (0.1–0.7 THz) are relatively large, leading to the smaller divergence angle for these frequency components in the $x > 0$ region and consequently smaller optimal x-direction displacement of the water line. At the optimal water line position for low-frequency components, the divergence angle of high-frequency components deviates from the optimal collection efficiency. However, when the water line gradually moves toward the negative direction, the collection efficiency of higher-frequency components increases, leading to the relative increase of higher-frequency components compared to those of lower frequency in the whole spectrum. Thus, the waveform variation shown in Fig. 7.8(c) and the difference in the THz spectrum shown in Fig. 7.8(d) can be explained.

A similar trend can be observed in the experiment using the 0.21-mm water line [16]. Fig. 7.9(a) shows the overall variation of the THz waveform when the water line moves along the x-axis in a two-dimensional manner. The consecutive waveform variation shown in Fig. 7.9(a) agrees with that in Fig. 7.8(c). The dependence of THz pulse energy on the x-position is plotted in Fig. 7.9(b), which shows a similar "two peaks" trend. Furthermore, Fig. 7.9(c) shows the dependence of the THz waveform on the z-position of the water line. The corresponding THz pulse energy variation is shown in Fig. 7.9(d). When the water line is scanned along the z-axis, the THz waveform experiences a temporal shift that almost linearly increases with the water line displacement. The temporal shift originates from the change

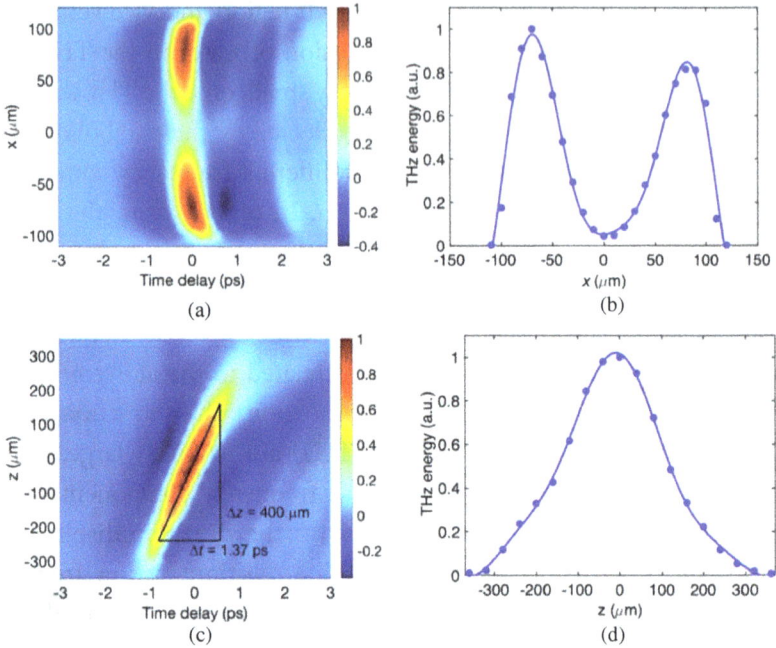

▲ Fig. 7.9. Dependence of THz wave electric field on (a) x- and (c) z-positions of water line. The color bar represents the THz wave amplitude. Corresponding THz energy dependence on (b) x- and (d) z-positions of water line. Reprinted from Ref. [16] with permission.

in optical path length caused by the variation in the relative position of the laser-induced plasma with respect to the center of the water line. However, the total range that the water line can move along the z-direction before the THz signal vanishes exceeds 600 μm, which is significantly longer than the diameter of the water line itself (210 μm). Such a phenomenon can be mainly attributed to the combined effect of the nonlinear propagation of the optical pump pulse along the laser beam path and the geometrical focusing of the cylindrical surface of the water line. The combined effect makes one laser focusing point (generating relatively stronger plasma) be sustained inside the water line in a wider range of the relative position between laser beam and water line, while another focal point that generates a weaker plasma remains in the air when the relative position is sufficiently long.

Based on the above analysis the phenomena of "two peaks" THz energy variation trend and waveform variation can be attributed to the combined influence of the THz wave absorption in water, refraction of the cylindrical water surface, and the abnormal dispersion of water in the frequency range from 0.1 to 3.0 THz. The cylindrical hyper-hemisphere water surface can enhance the collecting efficiency by converging the generated THz wave in the lateral direction, resulting in a 4.4-times enhancement in THz energy compared to the case of forward emission for the 110-μm water line [14]. Furthermore, the convergence effect of cylindrical hyper-hemisphere water surface occurs in water lines with different diameters. We repeat the above experiment using different water lines, and the results are shown in Fig. 7.10. The abscissa x/R (R: radius of the water line) is a dimensionless parameter to describe the x-direction displacement of water lines. The "two peaks" trend occurs in all these results using different water lines. However, the peak value in the $x > 0$ region is lower in comparison to that in the $x < 0$ region for water lines

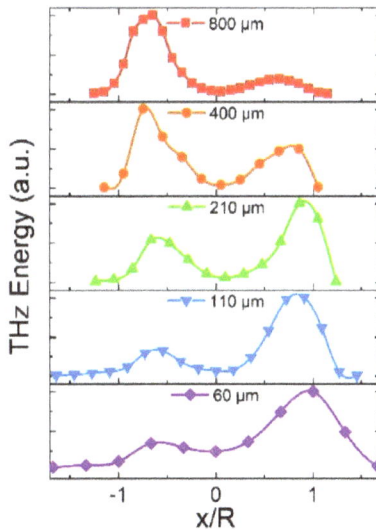

▲ Fig. 7.10. The dependence of THz energy (normalized to the maximal value for each curve) on the x-direction displacement for water lines with different diameters when the detection angle is 90°. R: radius of the water line. Reprinted from Ref. [14] with permission.

with larger diameters, since the absorption in water increases with the diameter of the water line. The variation of the peak value when $x > 0$ using different water lines verifies our qualitative explanation of the trade-off between THz absorption in water and the divergence angle of the THz wave. The experimental results and physical explanation also imply that the cost of the convergence effect is longer optical path in a liquid medium. In liquid water, the absorption loss makes the enhancement from the convergence effect less pronounced. Therefore, the enhancement from the convergence effect of the cylindrical water surface is more remarkable for thinner water lines. If liquid water used in our experiments is replaced by other liquids with lower absorption coefficients in the THz band, the enhancement effect will be even more remarkable due to the decrease of THz absorption loss during propagation.

7.6 Angular Distribution of THz Radiation

Three factors, the dipole orientation, the THz reflection and refraction on the water-air interface, as well as the absorption inside the water line, play important roles in the angular distribution of the emitted THz wave. In this section, we will introduce the angular distribution of the THz wave emitted from the water line as the water line position is optimized for lateral emission, and located at the center of the laser beam.

In order to measure the angular distribution of THz radiation, the collecting optical components and the detector (i.e., the detection system) are all installed on a platform that can be rotated around the water line. The detection system is rotated around the water line from 15° to 165° (limited by the finite space) with respect to the laser beam axis to measure the angular distribution when the THz signal is optimized at a detection angle of 90° (i.e., lateral emission), corresponding to position A in Fig. 7.8(a). These measured results on the THz emission distribution are shown as the black squares in Fig. 7.11, which indicate that the lateral emission is higher than forward emission while the largest THz energy can be detected at the angle of about 120° with respect to the laser beam axis.

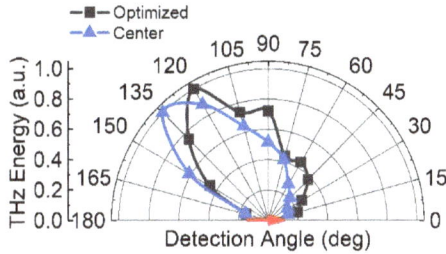

▲ Fig. 7.11.   Angular distribution of THz radiation from the water line when the laser focus is optimized for lateral detection (black squares) or set at the center of the water line (blue triangles). The red arrow points to the laser propagation direction.

However, the THz emission pattern may be distorted by the absorption in water and the refraction on the water-air surface when the water line position is optimized for detection at a certain angle. In order to give a nearly undistorted THz wave emission pattern from laser-induced plasma in liquid water, the angular distribution of THz radiation is measured when the laser focus is set at the center of the water line. Thanks to the circular transverse section of the water line, the THz wave emitted from the center has an equal optical path in water and is hardly affected by refraction on the water-air surface. As the laser-induced plasma is formed at the center of the water line, the distribution of the THz wave emitted by the plasma experiences as little distortion as possible after THz radiation is coupled out of the water line. The results are plotted as blue triangles in Fig. 7.11. The maximal THz radiation roughly falls between 120° and 150° with respect to the laser beam axis. Although the THz energy data point measured at 0° is absent in our results, based on the overall trend of the angular distribution we can draw a conclusion that the THz energy measured from lateral directions is larger than that measured in the forward direction.

Unexpectedly, the experimental result does not show a typical dipole emission pattern which has maximal THz radiation at 90° and is symmetrical with 90° as the axis. The discrepancy between the experimental result and the typical dipole emission pattern is probably due to the large absorption coefficient of the THz wave in the laser propagation direction.

The velocity of near-infrared light is higher than that of the THz wave in liquid water. After the THz wave is generated from laser-induced plasma, the defocused and scattered 800-nm laser pulse propagates in the remaining half of the water line and stimulates electrons from water, leaving a partially ionized zone. The absorption of the THz wave in this zone is higher due to the higher electron density. As a result, more THz wave is absorbed when it propagates through this trace, leading to the asymmetrical dipole-like emission pattern as shown in Fig. 7.11.

7.7 Pump Pulse Energy Dependence

The pump pulse energy dependence of THz radiation generated from laser-ionized liquid water has been demonstrated using flat and cylindrical water flows. A linear relationship between THz energy and pump pulse energy was obtained from both the free-flowing water film [2] and the water jet [7], as shown in Figs. 7.12(a) and (b), respectively. However, the authors believed that a linear dependence rather than a quadratic one is mainly caused by the plasma position displacement. In a follow-up work, a quasi-quadratic dependence between THz energy and pump pulse energy was demonstrated with a water jet scheme [5] as Fig. 7.12(c) shows. For the water line scheme, the linear scaling of the THz field strength (quadratic scaling of the THz energy) with the laser energy was verified by experimental and particles-in-cell (PIC) simulation results shown in Fig. 7.12(d) [10]. Although the influence of plasma position displacement was noticed in previous works, no effective solution has been demonstrated to reduce its impact. And no saturation was observed in either the water jet scheme or the water line scheme previously.

In order to investigate the pump pulse energy dependence for the forward emission scheme, we employ a circular variable metallic neutral density filter to control the incident laser pulse energy, and a water line with 0.20-mm diameter is used to generate the THz wave. Firstly, we optimize THz peak amplitude by changing the water line position when the pump pulse energy is set at 0.4 mJ. The generated THz peak amplitude decreases

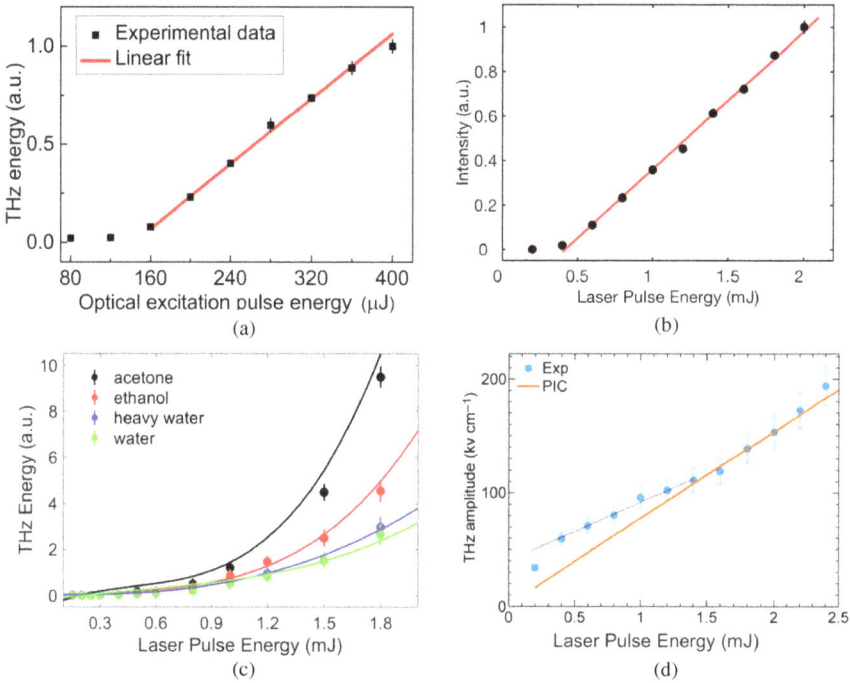

▲ Fig. 7.12.   Dependence of THz energy/intensity on pump pulse energy for (a) water film [2] and (b) [7], (c) [5] liquid jet scheme, respectively. (d) Dependence of the THz amplitude on pump pulse energy for 0.2-mm water line. Reprinted from Ref. [10] with permission.

when we simply increase incident pulse energy, as shown by the red dots in Fig. 7.13(a). However, if we optimize the water line position under 2.8 mJ pump pulse energy, the THz peak amplitude decreases when the pump pulse energy is gradually attenuated from 2.8 to 0.4 mJ, as shown by the blue dots in Fig. 7.13(a). These two trends are completely inconsistent. Since the water line scheme is extremely sensitive to the plasma position [11], the above discrepancy can be attributed to plasma displacement when the THz signal is optimized at different pump energies.

Due to the self-focusing effect in both air and liquid water, variable incident pulse energy leads to EFL change. Fig. 7.13(b) schematically shows the displacement of the laser focal point with respect to the water

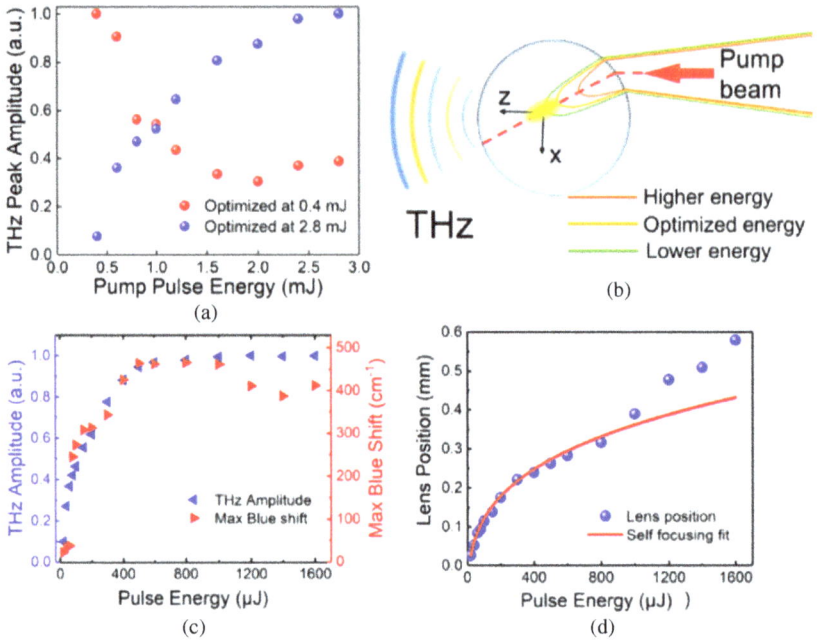

▲ Fig. 7.13.   (a) Normalized THz peak amplitude as a function of the incident pulse energy. Red dot: water line position optimized under 0.4 mJ. Blue dot: water line position optimized under 2.8 mJ. (b) Schematic diagram of the relative position between the water line and the laser focal point under different pump pulse energies. (c) Blue triangles: The THz amplitude dependence on pump pulse energy. Red triangles: The maximum blueshift of the broadened spectra as a function of the pump energy. (d) The optimal lens position as a function of pump pulse energy. Blue dots: experimental results. Red line: theoretical fit of the data using self-focusing effect. (a) and (b) are adapted from Ref. [13] with permission. (c) and (d) are reprinted with permission from Chen Y., He Y., Zhang Y., Tian Z. & Dai J. (2021). Saturation effect of THz emission from laser induced plasma in liquid water line, Frontiers in Optics + Laser Science 2021, paper JW7A.64.

line as the laser pulse energy varies. The yellow solid line defines the laser beam path when the water line position is optimized for the highest THz peak amplitude. The red dashed line represents the optical axis of the laser beam. When the laser pulse is attenuated to lower pulse energy, the focal point moves forward along the optical axis, and vice versa. On the other hand, the THz wave coupled out of the water line is highly dependent on the relative position between the water line and the laser focal point (i.e., the plasma position). The plasma displacement caused by the self-focusing effect has a significant influence on the efficiency of THz wave coupling out of the cylindrical surface of the water line, considering that the absorption of water and the refraction on the water-air interface will be different as the plasma position varies. The variation trend of THz peak amplitude will be unpredictable if we simply change the incident pulse energy without further optimizing the relative position between the water line and the laser focal point. To obtain the intrinsic pump pulse energy dependence, we move the focusing lens along the z-direction to compensate for the focal point displacement caused by the self-focusing effect in both air and inside the liquid water sample. In this condition, the variation trend of THz peak amplitude is independent of the initial pump pulse energy. As shown by the blue triangles in Fig. 7.13(c), THz peak amplitude increases rapidly with lower pulse energy and tends to be saturated at higher pulse energy. Different from the linear result in [10], saturation effect can be observed when pump pulse energy is higher than saturating pulse energy. This phenomenon can be attributed to the intensity clamping effect and consequently plasma density saturation in liquid water. The intensity clamping effect limits maximum laser intensity inside the plasma, and subsequently plasma density [18], thus limiting the peak amplitude of generated THz wave.

In order to prove that intensity clamping in plasma is the origin of THz emission saturation, a measurement of laser intensity inside the plasma is necessary. However, due to the high laser intensity inside plasma, precise measurement of the intensity distribution is difficult. An alternative and independent way to determine relative intensity inside the plasma can be achieved by measuring the pump pulse spectral broadening. In the work done by Prof. S. L. Chin's group [19], the spectrum broadening for laser pulse in water can be written as:

$$\Delta\omega(z, t) = -aI_0 \frac{\partial f(t)}{\partial t} + bI_0^m f^m(t) \tag{7.1}$$

where $f(t)$ is the temporal profile of the laser pulse, and I_0 is the peak intensity. The first term on the right represents a frequency shift due to

the nonlinear refractive index of water, and the second term is associated with plasma generation via multiphoton excitation ($m = 5$ for 1.55 eV photon energy) which causes a blueshift of the entire spectrum only. Thus, the frequency blueshift $\Delta\omega_+$ is expected to be dominated by the second term in the equation when the peak intensity is getting higher. In this case, the frequency blueshift is directly in correlation with the peak intensity: $\Delta\omega_{Max+} \propto I_0^m$. Since the $\Delta\omega_{Max+}$ essentially depends on the peak laser intensity, the saturation of $\Delta\omega_{Max+}$ occurs when the laser intensity inside plasma is saturated. Because the absorption near 800 nm is very low (absorption coefficient is much less than 0.1/cm [20]), we concluded that liquid water absorption near 800 nm can be neglected and will not affect the spectrum-broadening saturation when the diameter of the water line is less than 1 mm. Thus, $\Delta\omega_{Max+}$ saturation at high laser pulse energies can therefore be used as an indication of the saturated peak intensity. To verify the peak intensity inside, the plasma in our experiment is saturated. A fiber-coupled spectrometer (Ocean Optics USB2000+) is employed to monitor the laser pulse spectrum as the laser pulse energy is increased. The probe of the spectrometer is placed along the x-axis after a short-pass filter and a collecting lens, as shown in Fig. 7.2(a). The spectrometer records the broadened spectrum under different pump laser pulse energy when the water line position is optimized to the maximum THz peak amplitude. A saturation effect of spectral broadening can be observed from the short wavelength side. As indicated by the red triangles in Fig. 7.13(c), $\Delta\omega_{Max+}$ increases rapidly at low laser energy and tends to be saturated at high laser energy. The result of the maximum blueshift agrees well with the saturation trend of the THz amplitude, which verifies that intensity clamping or plasma density saturation is the main mechanism for THz peak electric field saturation since the generated THz peak amplitude also highly depends on the laser intensity or plasma density.

In order to give a clearer picture of the impact of the self-focusing effect on the focal point displacement, we record the focusing lens position after compensating for the focal point displacement under different pump energy for the water line with a 0.20-mm diameter as shown by the blue dots in Fig. 7.13(d). Also, we carry out a simplified calculation on the self-focusing-induced focal point displacement. In the calculation, the lens is calculated to be at different positions along the z-axis to counteract the self-focusing effect in water and air, thus keeping the focal point in water fixed under different pulse energy. The formula we used to trace the beam propagation is shown below:

$$\frac{a^2}{a_0^2} = \left(1 - \frac{p}{p_{cr}}\right)\frac{2z^2}{k^2 a_0^4} + \left[\frac{da}{dz}\bigg|_0 \cdot \frac{z}{a_0} + 1\right]^2 \tag{7.2}$$

where a is the beam radius, a_0 is the initial beam radius, p is the power of the laser pulse, p_{cr} is the critical power of the self-focusing effect, z is the laser propagation distance along the z-axis, and k is the wave vector of the laser beam. Since the critical power and wave vector are different in water and air, the calculation is carried out in two steps. First, we set the focal point in water as constant and trace the beam radius back to the incident plane. And then we use the beam parameters at the incident plane to trace the beam back to the focusing lens. Thereby, we calculate the lens position under different pump energies. The red line in Fig. 7.13(d) represents the calculation results and the blue dots represent the lens position obtained experimentally. The calculation roughly agrees with the experimental results. Under the optimal pulse duration of a 0.20-mm water line, the critical energy of self-focusing in the air is about 1.3 mJ while the one in water is three orders of magnitude smaller. When the pulse energy is less than 1.3 mJ, the lens needs to compensate only for the self-focusing effect in the water. When the pulse energy is higher than 1.3 mJ, the focal point displacement is a combination of self-focusing effects in both air and water.

It is noteworthy that the saturating pulse energies are different for water lines with different diameters. The saturating pulse energy increases with the water line diameter, as indicated by the vertical dashed lines in Fig. 7.14. The saturating pulse energy is about 0.4 mJ for 0.11 mm, 0.8 mJ for 0.20 mm, and 1.0 mJ for 0.39 mm water lines, respectively. The pump pulse energy dependence of water lines of different diameters was tested under

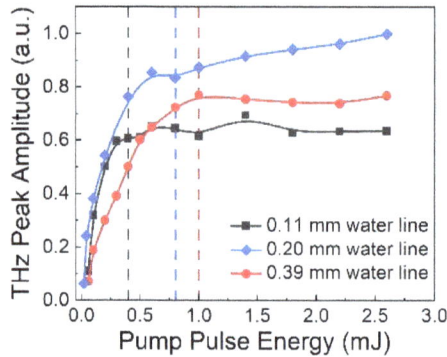

▲ Fig. 7.14.   Dependence of THz amplitude on pump pulse energy for water lines with 0.11-mm, 0.20-mm, and 0.39-mm diameters. Reprinted from Ref. [13] with permission.

their optimized pulse durations, as shown in Fig. 7.3(b). The optimal pulse duration is longer for a water line with larger diameter. Since the saturation effect is attributed to intensity clamping, the increase of saturating pulse energy is partially associated with the increased optimal pulse durations as the intensity decreases with pulse duration for fixed pulse energy.

The THz pulse energy and the optical-to-THz conversion are limited by the saturation of THz amplitude as pulse energy is higher than saturating pulse energy. The THz pulse generated from laser-ionized liquid water is always accompanied by incoherent infrared radiation. In this case, both the incoherent components and the coherent THz pulses will be detected by incoherent measurements (such as the Golay cell), leading to overestimation of the THz pulse energy. But these incoherent components cannot be detected by EO sampling. We have tried to measure the THz energy using a Golay cell (Tydex). When only two slices of high-resistivity silicon wafers are used as filters, the signal from the Golay cell is relatively large. However, after carefully filtering out the infrared components, the signal from the Golay cell is under the noise level. This phenomenon reveals that most of the signal detected by the Golay cell is from the infrared components. To give an approximate estimate of the coherent THz energy, we compare the THz signal from laser-ionized liquid water with the one from two-color laser-induced air plasma using EO sampling measurement. The 1.7-mJ

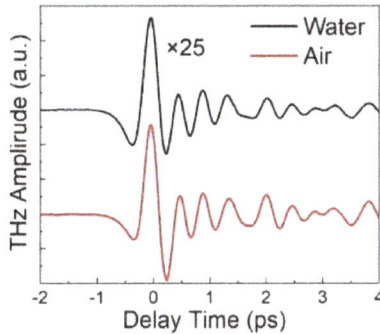

▲ Fig. 7.15. Comparison of the THz waveform from water line (black) and two-color laser-induced air plasma (red).

energy, 800-nm central wavelength laser pulses with 100-fs pulse duration are used in the two-color THz generation scheme. A 100-μm thickness β-barium borate crystal is used to generate the second harmonics and then the two-color pulses are focused by a two-inch EFL parabolic mirror. The THz pulse energy from the abovementioned two-color THz generation setup is on the sub-μJ scale, and the conversion efficiency of the two-color scheme is about 10^{-4}. As Fig. 7.15 shows, the peak electric field of the THz pulse from two-color air plasma is 25 times higher than that from laser-ionized water. And so, the THz pulse energy is about 660 times larger. By comparison of the EO sampling signals, we believe the THz pulse energy detected from laser-ionized liquid water in the forward direction is on the sub-nJ scale. The laser energy we use to generate the THz signal from the water line is 0.8 mJ. As a result, the conversion efficiency from near-infrared (NIR) to THz of laser-ionized water can be calculated to be on the 10^{-6} scale. However, the THz generation saturation effect in ionized water reduces the conversion efficiency from IR to THz. Moreover, the directionality of THz emission from ionized water is significantly worse than that of the two-color scheme. Only about one-tenth of the THz energy is collected in our experiment (calculated by the angular distribution shown in Fig. 7.11), leading to the underestimation of THz energy and consequently conversion efficiency. Overall, we believe the total conversion efficiency of THz generated from laser-ionized liquid water should be on

the 10^{-5} scale. Different approaches, such as asymmetry field excitation [21] and double pump technique [22–24], have been proposed to improve the optical-to-THz conversion efficiency at the cost of complexity. In addition, collecting sideway emission THz wave from laser-irradiated water line can directly improve the collection efficiency, thereby increasing the optical-to-THz conversion efficiency without increasing the complexity of the optical setup. With these approaches, there is a possibility to increase the conversion efficiency to the order of 10^{-4}.

7.8 Conclusion and Discussion

In conclusion, we systematically introduce the characteristics of THz wave emitted from water lines under the excitation of sub-picosecond laser pulses. The water circulation system and the experimental setup are demonstrated in detail. The experimental results on the water line diameter dependence show that the optimal pulse duration increases with the diameter of the water line. The water line with a diameter of about 0.10–0.20 mm generates the highest THz wave peak amplitude and a THz frequency redshift is observed when the water line diameter increases.

In investigations on the impact of relative position of the water line for forward emission THz wave, the THz amplitude first increases and then decreases with deviation of the water line from the laser axis. As the water line oppositely deviates from the laser axis by the same distance, the waveforms of the THz radiation are reversed while the absolute values of the THz field are nearly identical. The optimal deviation of the water line from the laser axis for the most efficient THz wave coupling is about 68% of the radius of the water line, independent of the actual diameter of the water line. The energy of lateral THz emission from the liquid water line is about 4.4 times that from the forward emission. The "two peaks" THz energy variation trend and THz waveform change as a function of the deviation of the water line from the laser axis are related to THz wave absorption, abnormal dispersion in water, and refraction on the water-air interface. The enhancement in the detected THz energy is attributed to the higher-efficiency lateral collection scheme and the convergence effect of the hyper-hemisphere cylindrical water surface. Such enhancement is more obvious for thinner water lines. It is noteworthy that the cylindrical water surface can only produce the convergence effect in one dimension and that utilization of liquid microdroplet may further increase the collection efficiency by converging the generated THz wave in two dimensions. Moreover, using a hyper-hemisphere HRFZ-Si lens can be an alternative way to increase the collecting efficiency.

The results on the angular distribution of coherent THz radiation indicate that the maximal THz radiation falls between 120° and 150°, and prove that collecting THz radiation in the lateral direction is a better choice to achieve higher THz collection efficiency in comparison to the forward collection scheme.

An intrinsic pump pulse energy dependence can be obtained by compensating the focal point displacement caused by the self-focusing effect. As the laser pulse energy increases, saturation in THz peak amplitude is observed and can be explained by intensity clamping and subsequently plasma density saturation inside the plasma. An independent measurement of laser spectrum broadening further verifies that the intensity clamping effect plays a major role in preventing THz peak amplitude from further increasing as the pump laser pulse energy increases. We estimate the energy of coherent THz pulse from the water line to be on the sub-nJ scale. The total generation efficiency of the coherent THz wave from laser-ionized liquid water in our experiment is on the 10^{-5} scale, which is an order of magnitude lower than that with the two-color laser-induced air plasma scheme. However, one should keep in mind that the detectable THz signal from liquid water lines is about two to three orders of magnitude lower than that from two-color laser-induced air plasmas, since the dipole-like emission pattern of the liquid water THz source causes only about 1–10% of the total THz radiation to be detected by regular THz time-domain spectroscopy systems, while THz radiation from two-color laser-induced air plasma is very directional and nearly 100% of the THz radiation can be detected.

We hope to contribute to a better understanding of THz wave generation and coupling-out processes in water lines with this chapter, and help to further optimize the collection efficiency of the generated THz radiation.

Acknowledgement

This work is supported by National Science Foundation of China (NSFC) (No. 62075157, 61875151), and National Key Research and Development Program of China (No. 2017YFA0701000).

References

1. Dey I., Jana K., Fedorov V. Y., Koulouklidis A. D., Mondal A., Shaikh M., Sarkar D., Lad A. D., Tzortzakis S., Couairon A. & Kumar G. R. (2017). Highly efficient broadband terahertz generation from ultrashort laser filamentation in liquids, Nature Communications, 8, pp. 1184.
2. Jin Q., E Y., Williams K., Dai J. & Zhang X.-C. (2017). Observation of broadband terahertz wave generation from liquid water, Applied Physics Letters, 111, pp. 071103.
3. Balakin A. V., Coutaz J.-L., Makarov V. A., Kotelnikov I. A., Peng Y., Solyankin P. M., Zhu Y. & Shkurinov A. P. (2019). Terahertz wave generation from liquid nitrogen, Photonics Research, 7, pp. 678–686.
4. Cao Y., E Y., Huang P. & Zhang X.-C. (2020). Broadband terahertz wave emission from liquid metal, Applied Physics Letters, 117, pp. 041107.
5. Tcypkin A. N., Ponomareva E. A., Putilin S. E., Smirnov S. V., Shtumpf S. A., Melnik M. V., E Y., Kozlov S. A. & Zhang X.-C. (2019). Flat liquid jet as a highly efficient source of terahertz radiation, Optics Express, 27, pp. 15485–15494.
6. Ismagilov A. O., Ponomareva E. A., Zhukova M. O., Putilin S. E., Nasedkin B. A. & Tcypkin A. N. (2021). Liquid jet-based broadband terahertz radiation source, Optical Engineering, 60, 8, pp. 082009.
7. E Y., Jin Q., Tcypkin A. & Zhang X.-C. (2018). Terahertz wave generation from liquid water films via laser-induced breakdown, Applied Physics Letters, 113, pp. 181103.

8. Huang H. H., Nagashima T., Hsu W. H., Juodkazis S. & Hatanaka K. (2018). Dual THz wave and X-ray generation from a water film under femtosecond laser excitation, Nanomaterials, 8, pp. 523.
9. Huang H. H., Juodkazis S., Gamaly E. G., Nagashima T., Yonezawa T. & Hatanaka K. (2022). Spatio-temporal control of THz emission, Communications Physics, 5, pp. 134.
10. Zhang L.-L., Wang W.-M., Wu T., Feng S.-J., Kang K., Zhang C.-L., Zhang Y., Li Y.-T., Sheng Z.-M. & Zhang X.-C. (2019). Strong terahertz radiation from a liquid-water line, Physical Review Applied, 12, pp. 014005.
11. Jin Q., E Y., Gao S. & Zhang X.-C. (2020). Preference of subpicosecond laser pulses for terahertz wave generation from liquids, Advanced Photonics, 2, pp. 015001.
12. Feng S., Dong L., Wu T., Tan Y., Zhang R., Zhang L., Zhang C. & Zhao Y. (2020). Terahertz wave emission from water lines, Chinese Optics Letters, 18, pp. 023202.
13. Chen Y., He Y., Zhang Y., Tian Z. & Dai J. (2021). Systematic investigation of terahertz wave generation from liquid water lines, Optics Express, 29, pp. 20477.
14. Chen Y., He Y., Tian Z. & Dai J. (2022). Lateral terahertz wave emission from laser induced plasma in liquid water line, Applied Physics Letters, 120, pp. 041101.
15. Wang H. Y. & Shen T. (2020). Unified theoretical model for both one- and two-color laser excitation of terahertz waves from a liquid, Applied Physics Letters, 117, pp. 131101.
16. Ling F., E Y., Fu S. & Zhang X.-C. (2021). Sideway terahertz emission from a flowing water line, in 46th International Conference on Infrared, Millimeter and Terahertz Waves (IRMMW-THz), pp. 1–2.
17. Wang T., Klarskov P. & Jepsen P. U. (2014). Ultrabroadband THz time-domain spectroscopy of a free-flowing water film, IEEE Transactions on Terahertz Science and Technology, 4, pp. 425–431.
18. Efimenko E. S., Malkov Y. A., Murzanev A. A. & Stepanov A. N. (2014). Femtosecond laser pulse-induced breakdown of a single water microdroplet, Journal of the Optical Society of America B, 31, pp. 534–541.
19. Liu W., Petit S., Becker A., Akozbek N., Bowden C. M. & Chin S. L. (2002). Intensity clamping of a femtosecond laser pulse in condensed matter, Optics Communications, 202, pp. 189–197.

20. Curcio J. A. & Petty C. C. (1951). The near infrared absorption spectrum of liquid water, Journal of the Optical Society of America, 41, pp. 302–304.
21. Jin Q., Dai J., E Y. & Zhang X.-C. (2019). Terahertz wave emission from a liquid water film under the excitation of asymmetric optical fields, Applied Physics Letters, 113, pp. 261101.
22. E Y., Jin Q. & Zhang X.-C. (2019). Enhancement of terahertz emission by a preformed plasma in liquid water, Applied Physics Letters, 115, pp. 101101.
23. Ponomareva E. A., Tcypkin A. N., Smirnov S. V., Putilin S. E., E Y., Kozlov S. A. & Zhang X.-C. (2019). Double-pump technique — one step closer towards efficient liquid-based THz sources, Optics Express, 27, pp. 32855–32862.
24. Ponomareva E. A., Ismagilov A. O., Putilin S. E., Tsypkin A. N., Kozlov S. A. & Zhang X.-C. (2021). Varying pre-plasma properties to boost terahertz wave generation in liquids, Communications Physics, 4, pp. 4.

© 2024 World Scientific Publishing Company
https://doi.org/10.1142/9789811265648_0008

Chapter 8

Terahertz Wave Generation from Unique Liquids

A. V. Balakin,[1] P. M. Solyankin,[1,2] A. P. Shkurinov,[1] Y. M. Zhu[3]

[1]*Lomonosov Moscow State University, Russia*
[2]*ILIT RAS-Branch of FSRC "Crystallography and Photonics",
Russian Academy of Sciences, Russia*
[3]*University of Shanghai for Science and Technology, China*

8.1 Introduction

There is a huge number of papers on terahertz (THz) spectroscopy of liquids and solutions as well as studies on properties of laser-induced nonlinear optical processes in liquid media. The most commonly used in experiments are polar (such as water) and nonpolar (such as acetone) solvents. These liquids are easily available and widely distributed, thus a large number of experimental methods have been developed to work with these liquids and to excite them with laser radiation. Therefore, the first experiments on THz generation were carried out in these widely known fluids ([44, 45], see also Chapter 2). There are a few unquestionable advantages of liquid as a laser-pumped THz radiation source compared with, say, solid. For example, a liquid after interaction with laser radiation can evaporate without forming splashes and without contaminating the experimental stand. Also, liquids make it possible to form a constantly renewable medium with a density close to the solid state. Finally, by selecting liquid media with an optimal combination of nonlinearity, conductivity, and absorption coefficients for pump and THz radiation, one can achieve the desired spectral-angular

characteristics of the signal and/or the maximum conversion coefficient. Because almost any substance can be in liquid state, that could allow one to find the optimal conditions for THz generation under laser excitation.

In this chapter, we use the term "unique liquids" to refer to fluids that are hard to maintain in liquid state with desired characteristics under normal conditions. We will discuss in detail cryogenic liquids like liquid nitrogen [1–4] and liquid metal media in the form of microdroplets and streams [5–7].

8.2 Terahertz Generation in Cryogenic Liquids

Cryogenic liquids, or cryoliquids, are usually defined as fluids with boiling point below 120°K [8]. The most frequently used cryoliquid is liquid nitrogen (LN). LN provides a number of benefits: it is a common and cheap material that can be produced from air, and it is inflammable — that is especially important in the case of laser excitation. After evaporation it produces gaseous nitrogen that is the main component of ambient air. Moreover, most of the experiments regarding THz generation in plasma were held in air, so a comparison between gaseous and liquid states of nitrogen can be done.

In [9] it was found that LN has high third-order nonlinearity. Being a nonpolar liquid, LN has moderate THz absorption [10]. Together with high optical transparency, it makes LN a prospective material for laser-to-THz conversion. To investigate THz properties of LN, we used a double-wall cryogenic cuvette of special design with z-cut quartz windows and an LN layer width of 19 mm. A common THz-time domain spectroscopy (TDS) system [33] provided us with information on the refraction index and absorption coefficient, which are depicted in Fig. 8.1. THz absorption has a clear peak near 1.7 THz with an absolute value less than 1 cm^{-1}, and refraction index was measured to be near 1.195 in the 0.2–1.7 THz range.

The main difficulty in the case of LN is holding the substance's temperature below the boiling point. Usually, double-wall cuvettes with a vacuum layer

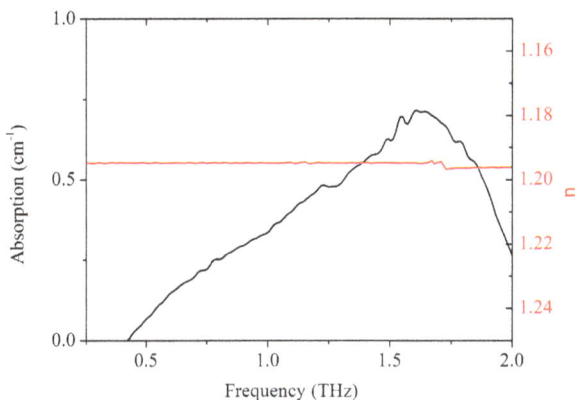

▲ Fig. 8.1. Absorption and refraction spectra of LN in the THz frequency range.

between the inner and outer vessels are used to minimize the heat flux. But in the case of high-energy laser pump and the presence of moderate THz absorption, it is hard to prevent the cuvette's windows from optical breakdown. So two alternative methods were demonstrated to exclude any windows and to work near the air-LN interface: to use free surface of liquid [1, 2] or thin jets [3, 4]. These two approaches are discussed below.

8.2.1 Experimental Setup with the Free Surface of LN

Like in air-based plasma, two types of excitation schemes were used: single- and double-color excitation. It is well known that the use of radiation on the doubled optical frequency can significantly increase the THz output from gaseous plasma due to asymmetrical ionization of the media. In the case of liquids, the growth of density leads to the increased divergence in time for optical pulses at fundamental and doubled frequencies due to dispersion, so additional experiments should be done to evaluate the efficiency of this excitation scheme.

To study laser interaction with LN using a region near the free LN-air surface, an experimental setup, depicted in Fig. 8.2, was proposed in [1]. Laser radiation was directed to the surface normal of the LN, and it was

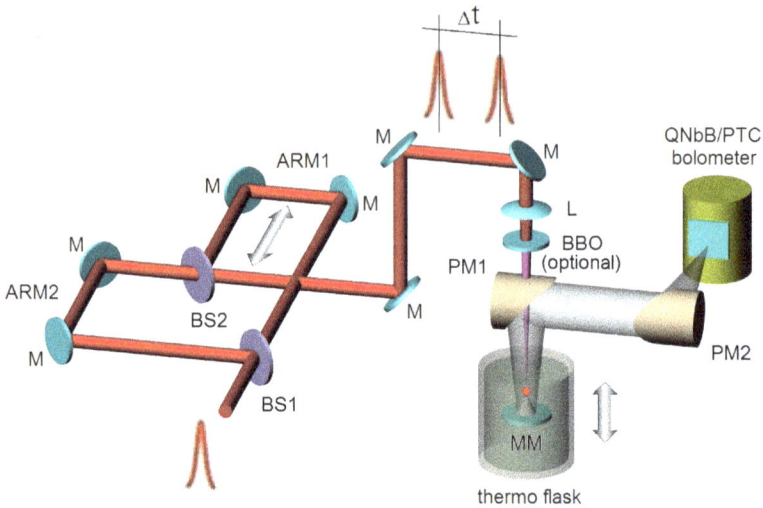

▲ Fig. 8.2.   Schematic of the experimental setup on THz generation in the vicinity of free surface of LN. BS1, BS2 — beam splitters 1:1, M — dielectric mirrors, MM — metallic mirror of special design, PM1, PM2 — off-axis parabolic mirrors, L — lens, BBO — β-barium borate nonlinear crystal. Reprinted from Ref. [1] with permission.

focused through the hole in PM1 inside the thermo flask. Generated THz radiation was reflected from the metallic mirror MM to PM1, and then into the sensitive element of the bolometer. Varying the vessel's altitude with the fixed position of all optical components, one can study THz generation in the air (gaseous nitrogen) or inside the LN volume. Nonlinear BBO crystal could be placed inside the laser beam to generate THz radiation in a double-color scheme. Regenerative amplifier Legend Elite Duo (Coherent, Inc.) with 1 kHz pulse repetition rate, 800 nm central wavelength, and 90 fs pulse duration was used as a source of laser pump radiation.

To analyze the features and dynamics of liquid media ionization, a double-pulse excitation scheme was implemented. Two codirectional laser pulses with a variable delay of up to a few picoseconds between them were formed using a Michelson-type interferometer. These pulses had the same linear polarization and equal energies up to 1.25 mJ per arm. The leading pulse caused the liquid ionization, while the second one enters excited media. Due to the indistinguishability of these pulses, the THz signal

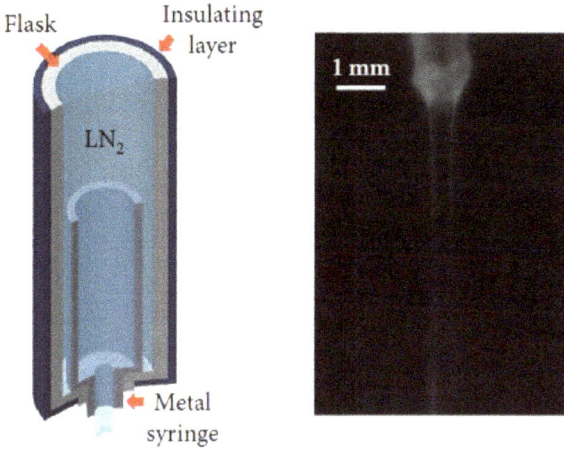

▲ Fig. 8.3.　Experimental realization of LN jet: apparatus cross-section (left) and resulting jet photograph (right). Reprinted from Ref. [4] with permission.

contained both components from each arm, and it was symmetric with respect to zero time delay.

8.2.2　Experimental Setup with Thin LN Jet

The second approach to deal with LN is to minimize the length of laser-matter interaction using thin jets. It was shown in [3] that it is possible to create a cryogenic LN jet in ambient air due to formation of a thin boundary layer of cold vaporized LN that prevents the main stream from boiling (so-called Leidenfrost effect [12]). Syringe needle coupled with the bottom of the thermo flask provided a gravity-driving LN jet, so no special pumps were used. Fig. 8.3 represents a cross-section of an elaborated jet apparatus construction and cross-section of the resulting stream. Thin jets tend to break up into droplets at a certain distance from the nozzle due to Plateau-Rayleigh instability. To maximize the stability of the jet properties, the LN-laser interaction point was chosen near the end of a needle with 400 mm inner diameter.

To achieve THz generation, the authors tightly focused single-color titanium:sapphire laser pulses with 370 fs duration and 0.4 mJ energy into

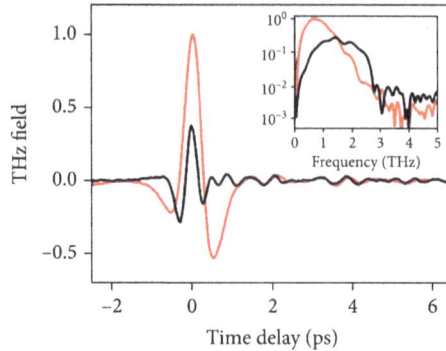

▲ Fig. 8.4.   Spectra of THz signal generated in 400 μm thick LN (black) and 210 μm thick water jets (red). Reprinted from Ref. [4] with permission.

the beam waist of 2.6 μm. THz signal was registered with the help of a TDS setup equipped with an electro-optic (EO) detector. Thus, the time profile and spectrum of THz pulses were measured. The thinness of the jet made it possible to neglect most of the propagation effects in the liquid.

8.2.3 THz Generation in LN Under Single-Color Excitation Regime

In experiments discussed in this paragraph, the LN medium is excited only with the fundamental laser frequency without addition of the second harmonic. The use of a thin jet together with TDS detection technique allowed the researchers to measure waveform and spectrum of the generated THz pulse, which were almost unaffected by propagation in liquid media (see Fig. 8.4) [4]. Peak amplitude of THz signal from the LN target was 2.5 times weaker compared to the peak value of the signal from the liquid water jet, but the spectrum was wider and in fact was limited by the EO detector feature. One should note that absorption of these two liquid jets in THz range differs around two orders of magnitude, so extra efforts to maximize THz output could be applied: one can vary the jet diameter, lateral shift of the focusing point and beam waist size to optimize the THz signal properties. For the given LN jet and beam waist parameters, maximal amplitude of the THz

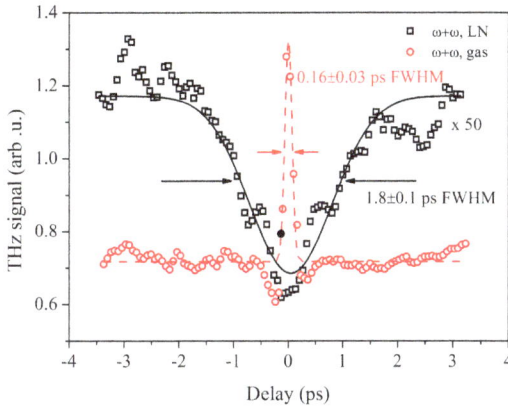

▲ Fig. 8.5.   THz yield versus time delay between the laser pulses in LN (black) and air (red) for single-color excitation scheme. Adapted from Ref. [1] with permission.

signal was registered at ±170 μm lateral shift from the center of the jet. It could be explained by taking into account changing of the incident angle of the laser radiation falling to the air-LN boundary and coupling with THz detector [13] or by features of the THz signal propagation and absorption in LN in the presence of optical radiation.

To study the dynamics occurring under optical excitation of LN, the double-pulse excitation scheme was applied (see Section 8.2.1). Fig. 8.5 represents the dependence of the THz signal power measured as a function from the delay between two optical pulses for both gaseous and liquid nitrogen media. The significant difference in behavior of these dependencies may undoubtedly be observed from the figure. In the case of the gaseous medium, a sharp single maximum at zero delay is observed. Its width is comparable with the laser pulse duration. Due to the nonlinear nature of the THz generation, this peak corresponds to a coherent sum of two laser pulses with the same polarization when they overlap in time. At the other time delays, i.e., outside of the overlapping region, the THz radiation generated from each arm of the interferometer sums incoherently. But in the case of the LN medium, one can clearly see a local minimum of the THz signal at zero time delay, and maximal values of the signal are reached at a few picosecond time shift, which is much higher than the laser pulse duration.

Such a phenomenon could be explained by taking into account formation of the ambipolar electrical field as a result of the ambipolar diffusion of charged particles in ionized liquid, as described in detail in Section 8.2.6.

8.2.4 THz Generation in LN Under Double-Color Excitation Regime

To study THz generation in LN under the double-color excitation regime, the experimental setup depicted in Fig. 8.2 was used. Both the fundamental laser frequency and its second harmonic, generated in 300 μm thickness BBO crystal placed after the focusing lens, were applied for excitation of the LN medium. As one can see from Fig. 8.6, for both gaseous and liquid nitrogen media, the dependence of THz signal versus delay between two excitation pulses demonstrates the same behavior: there is a sharp single peak near the zero time delay, and the width of the peak is comparable with the laser pulse duration.

Fig. 8.7 represents the power of THz radiation depending on the position **h** of the laser beam waist relative to the surface of the LN for a double-color excitation scheme. Negative values of beam waist position **h**

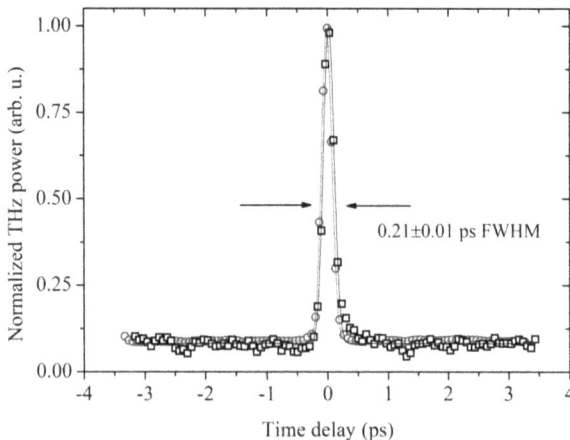

▲ Fig. 8.6.   THz yield versus time delay between the laser pulses in LN (black) and air (red) for double-color excitation scheme.

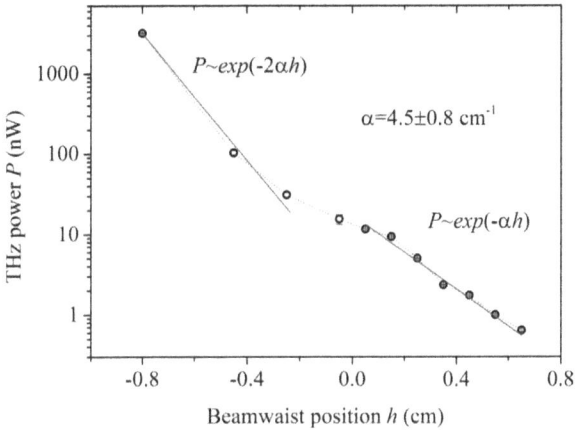

▲ Fig. 8.7.   THz power versus beam waist position relative to the surface of LN for double-color excitation scheme.

correspond to THz generation above the LN surface (i.e., in air), while positive ones correspond to the beam waist position inside the LN. In both cases one can see the attenuation, caused by THz absorption in the media. Total THz path L in the LN could be expressed via the beam waist position h as: $L = 2 \cdot (h + 0.8 \text{ cm})$ for $h < 0$, and $L = h + 1.6$ cm for $h > 0$, where 0.8 cm corresponds to the fixed distance between the flat mirror MM and the focal point (see Fig. 8.2). According to the Bouguer–Beer–Lambert law, the THz power scales as $P \sim \exp(-\alpha L)$, where α is the absorption coefficient. If we neglect the optical losses in LN, the experimental curve will take the form of two exponential decays with a smoothed shift around the LN surface level. Using experimental data from Fig. 8.7, it is possible to estimate the absorption coefficient of THz radiation in LN: $\alpha = 4.5 \pm 0.8$ cm^{-1}. This value is higher than measured in Fig. 8.1, and the difference could be attributed to boiling or nonlinear processes inside the LN media. The shift between two exponential curves near the surface of LN corresponds to the different optical-to-THz conversion efficiencies in air and LN. Considering the experimental errors, it is possible to estimate an increase in the efficiency of THz generation in LN by a factor of 4–10 relative to the gaseous medium. For more accurate calculations, one should take into account the contribution of the additional nonlinearity of the LN surface and the losses

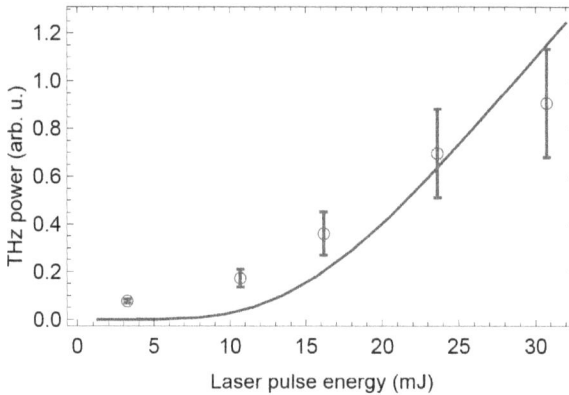

▲ Fig. 8.8.   Power of THz signal versus laser pulse energy for the double-color excitation scheme. Reprinted from Ref. [1] with permission.

due to reflection (and scattering) of femtosecond laser radiation passing through the media interface.

Fig. 8.8 represents the dependence of registered THz yield as a function of pump laser energy in the double-color excitation scheme. Solid line corresponds to the theoretical fit in the framework of the transient photocurrent model (see Section 8.2.6.1).

8.2.5 Nonlinear Effects During Propagation of THz Pulse Through LN in the Presence of Laser Pump Radiation

It was found in [2] that high-intensity laser pump radiation, used to generate THz waves, can affect polarization properties of generated THz pulses during their joint propagation through the LN medium. It is well known that LN is an isotropic centrosymmetric media under normal conditions, so the change in polarization state of the THz wave could be attributed only to the odd-order nonlinear effects. Such type of nonlinear process in the optical region is well known as a self-rotation of the polarization ellipse [11]. This phenomenon is observed for elliptically polarized radiation and increases with the growth of the polarization ellipticity degree and optical radiation intensity, and it can be explained in terms of χ^3 nonlinearity.

The peak intensity of currently available THz radiation is quite low, so the direct analogues of the effect detected by Terhune for optical waves in [11] is rather small to be revealed for THz waves. But the Terhune-like transformation of the THz polarization ellipse was observed in [2] by joint propagation of three waves (two optical and one THz) in LN. Double-color scheme was applied for the generation of THz wave in these experiments. As was described in [1, 2], the initial optical pulse passed through a BBO crystal to generate second harmonic radiation. Due to birefringence in the BBO crystal, the optical wave at fundamental frequency got an essential ellipticity. The experimental setup, depicted in Fig. 8.2, allowed one to generate THz radiation in air plasma directly above the LN-air boundary, and then to study transformation of the THz wave polarization after passing through the LN layer together with the initial laser radiation. By lifting up or putting down the Dewar vessel, one could register the polarization state of the THz radiation in the presence of the LN layer or without it. Acquired experimental data are depicted in Fig. 8.9. As one may clearly see, the main axis of THz wave polarization ellipse rotates up to 14°, which is accompanied with change of its ellipticity after passing through

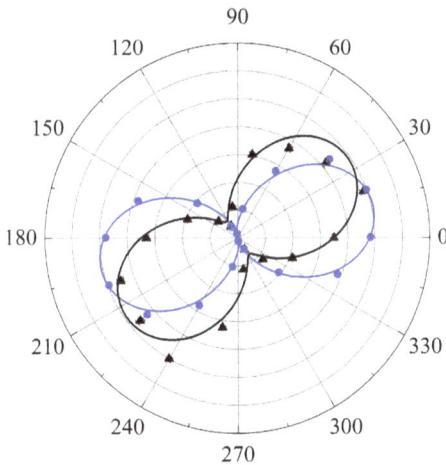

▲ Fig. 8.9.   Change of polarization state of the THz radiation after propagation through LN in the presence of laser pump radiation. Blue curve corresponds to the initial polarization of THz wave (without LN layer), while black curve corresponds to polarization of THz wave having passed through a 13 mm layer of LN. Reprinted from Ref. [2] with permission.

a 13 mm layer of LN. Dispersion of the LN medium leads to different refractive indices for THz (n_{THz} ~ 1.195), fundamental laser wave (n_{800} ~ 1.196) and second laser harmonic (n_{400} ~ 1.202) that limit the effective overlapping length of co-propagating pulses. Four-wave mixing theory-based approach was successfully used to describe this effect (see [2] and Section 8.2.6.2). These experimental observations emphasize that various nonlinear processes occurring during the THz radiation propagation as well as additional absorption of THz radiation in the presence of residual laser pump radiation should be taken into account when elaborating the theoretical model of propagation in the liquid media.

8.2.6 Theoretical Models for THz Generation in LN

8.2.6.1 Transient Photocurrent Model

One of the possible mechanisms of THz generation in the case of double-color excitation of LN is a transient photocurrent model (also known as asymmetric ionization) proposed in [37] for description of THz generation in laser-induced air plasma. If one assumes a double-color laser field as a plane wave with only one nonzero component along the x-axis, $E_x = E_\omega \cos(\omega t) + E_{2\omega} \cos(2\omega t + \psi)$, substitute this expression into the equation of motion for free electrons in collisionless approximation, then after integration over one laser period ($2\pi/\omega$) the following drift component of the electron velocity, i.e., mean value of single-electron velocity, could be easily evaluated as $V_x(t_0) = \frac{eE_\omega}{2m_e\omega}\left(-\frac{E_{2\omega}}{E_\omega}\sin(2\omega t_0 + \varphi) - 2\sin(\omega t_0)\right)$. Here it is assumed that free electron is born through photoionization at time t_0. Next, one can get the production of the rectified current, which is proportional to the THz emission, as $\frac{\partial}{\partial t}\vec{J}(t) \approx e\vec{V}(t)n_f w(E(t))$ (see [38]), where $w(E(t))$ is the probability of photoionization. So one can see that the rectified current is nonzero only in the presence of the second harmonic of optical radiation. Following [1], one can calculate the dependence of total THz signal on the power of pump laser radiation. The corresponding fit curve is presented in Fig. 8.8 for approximation of experimental data. This approximation curve reflects the trend quite well, and some disagreement with measured values could be explained by the use of collisionless assumption in the framework of the developed transient photocurrent model. As would be discussed in the next section, the gradual decrease of the mean free path of electrons should be taken into account in the case of liquids.

8.2.6.2 Four-Wave Mixing in the Case of Ambipolar Diffusion

In the framework of four-wave mixing (FWM) theory [34], one can write the expression of third-order nonlinear polarization of the medium as:

$$P_i^{(3)}\left(\omega_q\right) = \chi_{ijkl}^{(3)}\left(\omega_q;\omega_m,\omega_n,\omega_r\right)E_j(\omega_m)E_k(\omega_n)E_k(\omega_n)E_l(\omega_r), \quad (8.1)$$

where $\omega_q = \omega_m + \omega_n + \omega_r$ and E is a complex amplitude of the wave at corresponding frequency. LN is an isotropic centrosymmetric medium, so the third-order susceptibility tensor could be simplified to $\chi_{ijkl}^{(3)} = A\delta_{ij}\delta_{kl} + B\delta_{ik}\delta_{jl} + C\delta_{il}\delta_{jk}$ due to the symmetry of the substance. According to the classical theory of nonlinear polarization [34], all three coefficients are equal:

$$A = B = C = \sum_{(mnr)} \frac{bn_f e^4/m_e^3}{G(\omega_q)G(\omega_m)G(\omega_n)G(\omega_r)}; \quad G(\omega) = \omega_0^2 - \omega^2 - 2i\gamma\omega \quad (8.2)$$

Thus, one can estimate single coefficient $\chi^{(3)} \approx \frac{e^4 n_f}{m_e^3 \omega_0^6 d^2}$. Here e and m_e are the charge and mass of the electron, respectively, n_f is density of a substance, ω_0 corresponds to atomic frequency, d is atomic dimension, and $n_f \sim d^{-3}$ for liquids. Therefore, the third-order susceptibility is proportional to the material density and rises dramatically in LN relative to the gaseous counterpart.

One can rewrite Eq. (8.1) for the case of polarization at THz frequency, assuming the double-color laser pump and the presence of a certain quasi-DC field in the medium:

$$\begin{aligned}
P_i^{(3)}(\Omega) = \; &\chi_{ijkl}^{(3)}(\Omega;-2\omega,\omega,\omega)E_j(2\omega)E_k(\omega)E_l(\omega) \\
&+ \chi_{ijkl}^{(3)}(\Omega;\Omega_{DC},\omega,-\omega)E_j(\Omega_{DC})E_k(\omega)E_l(\omega) \\
&+ \chi_{ijkl}^{(3)}(\Omega;\Omega_{DC},2\omega,-2\omega)E_j(\Omega_{DC})E_k(2\omega)E_l(2\omega).
\end{aligned} \quad (8.3)$$

Here, Ω is THz frequency, Ω_{DC} corresponds to the quasi-DC field, and ω and 2ω are the fundamental frequency of the pump laser radiation and its second harmonic, respectively (see Fig. 8.2). The last two terms of this equation govern the THz polarization transformation in the presence of strong laser pulses (see Section 8.2.5). In this case Ω_{DC} should be replaced with Ω. In the case of single-color excitation, only the second term should be taken into account.

Now we should make some remarks on the ionization process in LN. The following estimations could be done for experimental conditions, listed in Section 8.2.1. The upper estimation of the emerged free electron concentration, made in the assumption of total absorption of laser radiation in the beam waist area, gives a value around $n_e \sim 10^{21}$ cm^{-3}, which is orders of magnitude less than the material density ($\sim 10^{23}$ cm^{-3}) [1]. Thus, the liquid is weakly ionized. Due to the high density of liquids in comparison with gases, mean free path of free electrons in liquids is shorter, and it could be estimated in the laser beam waist according to $\lambda_e = \frac{1}{n f \sigma}$, where σ is the cross-section of electron scattering. With the estimation of $\sigma = 10^{-16}$ cm^2, one can get $\lambda_e = 0.1$ μm. This value is less than the beam waist size $a > 8$ μm. Therefore, the motion of free electrons in the liquid seems to be diffusion. Diffusion tries to smooth the initial spatial distribution of laser-born electrons. In plasma this diffusion is accompanied with the occurrence of radial quasi-DC electric field, called ambipolar field, that leads to equalization of radial fluxes of electrons and ions. According to [35], one can estimate that field as $E_{amb} = \frac{T}{e} \frac{\nabla n_e}{n_e}$, where T is the temperature of electrons. With the rough estimation of $T = 100$ eV and initial gradient located at beam waist size, one can evaluate E_{amb} up to 20 kV, which is higher than the external DC field in air plasma experiments [36].

In Eq. (8.3), the first term is proportional only to the laser field, while the second and third terms also include the ambipolar field. It is obvious that $E_{n\omega} \gg E_{amb}$. On the other hand, according to Eq. (8.2), $\frac{\chi^{(3)}_{ijkl}(\Omega,\Omega,n\omega,-n\omega)}{\chi^{(3)}_{ijkl}(\Omega,-2\omega,\omega,\omega)} \sim \frac{n\omega^2}{\Omega} \gg 1$.

Thereby, the contribution of ambipolar field can be significant for THz generation.

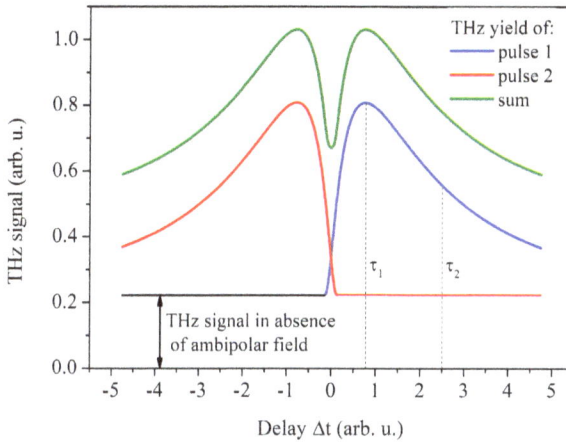

▲ Fig. 8.10.   Schematic representation of resulting THz yield from two pulses with variable delay. Reprinted from Ref. [1] with permission.

The role of ambipolar field in the process of THz generation in LN under single-color excitation regime is shown schematically in Fig. 8.10. Using a phenomenological approach with the single-pulse excitation, $\chi^{(2)}$ nonlinearity gives the primary contribution to the THz signal, but this process is ineffective in centrosymmetric media. As was discussed previously, during laser ionization of liquids, a quasi-DC ambipolar field emerges. In the presence of the DC field, a more effective process based on $\chi^{(3)}$ nonlinearity is taking place (see Eq. (3)). In the double-pulse excitation scheme, the second pulse is affected by the ambipolar field that was created by the first pulse. The bolometric detector registers the sum of THz signals generated from both first and second pump laser pulses, and due to equal polarization and energy, these pulses are indistinguishable. That leads to the symmetry of graph wings in Fig. 8.10. Assuming that the ambipolar field is characterized by formation time τ_1 and time of decay τ_2, and taking the formation time as ~1 ps, one can explain the presence of local minimum in the experimentally measured curve around zero time delay with width of ~2 ps (see Figs. 8.5 and 8.10). Indeed, in the central part of the graph the first laser pulse generates THz signal via a $\chi^{(2)}$ process, and starts the formation of the ambipolar field. The second

laser pulse comes into the excited media, where the ambipolar field did not reach its maximal value yet, so according to Eq. (8.3), the THz signal is growing with increasing delay between laser pulses. After time τ_1 the ambipolar field starts to decay, and the THz signal decreases. This process is illustrated in Fig. 8.10.

8.2.7 Summary

LN could be considered a prospective medium for THz generation under laser irradiation. It can provide debris-free interaction with the pump radiation and wide generation spectrum. Conversion efficiency for the LN medium exceeds this value in air for both single- and double-color excitation schemes. On the other hand, the experimental setup becomes more complicated, and additional optimizations should be done to reach the same THz signal amplitude as the usual water jet. Transient photocurrent along with the FWM and the ambipolar field formation should be taken into account on the development of THz generation model in liquids.

8.3 Terahertz Generation in Liquid Metals

Metals are widely used for nonlinear optics applications due to the high material density and thus increased nonlinear properties. And high concentration of electrons in metals makes metallic media attractive for applications aimed at getting effective THz generation. Ionization threshold for metals is relatively low, so lower laser intensities are needed. A large number of works on THz generation on metal surfaces [14], thin foils [15, 16] and wires [17, 18] were published. THz radiation generated jointly with X-ray [19] and bunches of electrons [20] had been observed under different excitation regimes. Up to multi-mJ energies in the THz pulses were registered [29]. Despite rather high optical-to-THz conversion efficiency in such solid-state sources, usually they suffer from laser-caused damage, and it compels one to mechanically move the targets.

To overcome this disadvantage and provide a renewable target for each laser pulse with comparable density, one can use liquid metal media. It could be realized in the form of constant-flow metallic jets [6], or a series of metallic droplets of microscopic size [21]. The second is widely used in UV lithography for forming a deep UV-pulsed radiation source [22]. These objects are recovered after each interaction with a laser pulse, and the quality of their surface — such as smoothness and geometry — is automatically maintained by the surface tension of the liquid.

From the point of view of practical realization, it is rather easier to deal with a continuous jet of liquid metal than with liquid metal droplets — there is no need to synchronize the droplets with laser pulses in time. On the other hand, the use of microdroplets allows one to excite sub-wavelength spherical targets as well as to decrease debris and obstacles in the process of THz beam generation. Besides, one can modify the shape of microdroplets with the use of laser pre-pulses [21], so various target geometries could be set. In this section, we will discuss the use of liquid microdroplets made of In:Sn alloy and liquid Ga jet as targets for laser-excited sources of THz radiation.

8.3.1 Single Liquid Microdroplet as Target for THz Generation

8.3.1.1 Experimental Setup

The scheme of the experimental setup for laser interaction with microdroplets and its photograph are depicted in Figs. 8.11 and 8.12, respectively. To measure angular distribution of the THz radiation generated from the microdroplet source, a special vacuum chamber housing on the base of 9.5 mm thick polypropylene (PP) tube with outer diameter of 179 mm was employed. PP possesses good transparency in the THz spectral range [23]. The chamber was equipped with three fixed position input/output ports for delivery of the excitation laser beam inside the vacuum chamber and registering the laser-microdroplet interaction process by shadowgraph technique (see Fig. 8.11). Such configuration of the ports provided focusing of laser radiation passing through the quartz window on the microdroplet, and monitoring of the droplet's size together with accuracy of the focusing, positioning and the simultaneity of laser

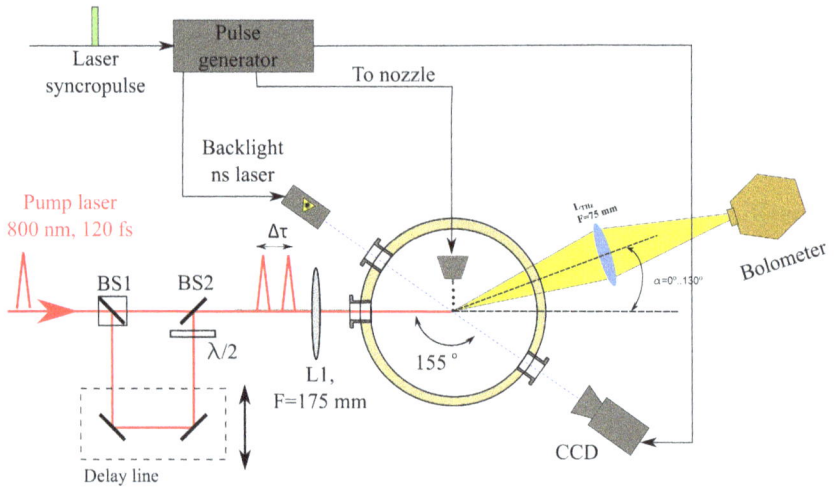

▲ Fig. 8.11.   Scheme of the experimental setup for THz generation in liquid metal micro-droplets. BS1 — Glan prism, BS2 — 1:1 beamsplitter, L1 — lens with $F = 175$ mm, CCD — digital camera, Bolometer — He-cooled bolometric detector.

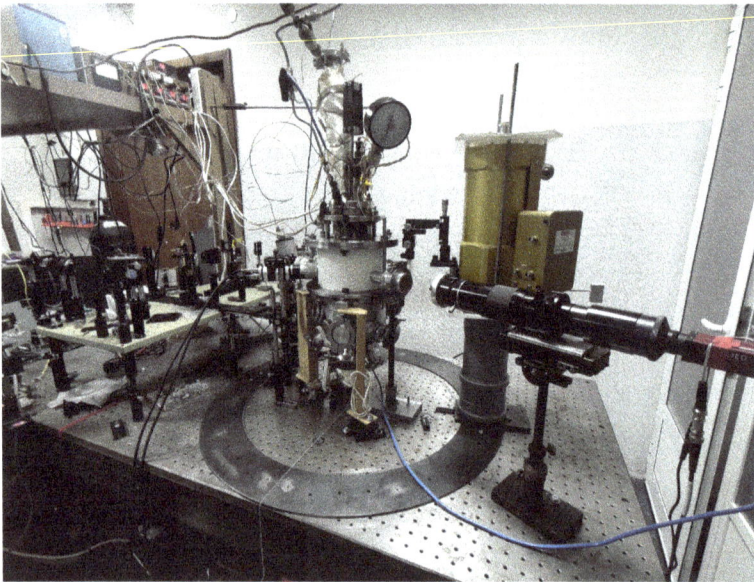

▲ Fig. 8.12.   Photograph of the experimental setup for THz generation in liquid metal microdroplets.

pulse and microdroplet arrival at the interaction point. To focus the laser radiation on the droplet target, a plano-convex lens L1 with focal length $F = 175$ mm was used, which provided a spot size of 15 μm full width at half maximum in the focal point.

A eutectic alloy of tin and indium (48% Sn, 52% In), which is in the liquid state at 140°C, was used as a material for the microdroplets. A sequence of freefalling single microdroplets with a diameter of 40–60 μm was formed with the help of high-temperature dispenser device MJ-SF-01 (MicroFab Technologies, Inc.) with the integrated piezoelectric actuator. The size of the droplets was set by varying the pressure value inside the chamber with liquid metal alloy and the profile form of managing voltage coming to the actuator. Laser-to-droplet alignment accuracy was adjusted in the horizontal plane by a micrometric screw on the lens translation stage, and in the vertical plane by the time delay between electrical pulses managing the droplet piezoelectric actuator and the laser synchronization pulse. Pressure in the vacuum chamber was kept at a level of 10^{-4} mBar to prevent oxidation of the metallic target. To monitor the droplet's size and positioning, time-resolved shadow photography method with 3 μm/pcs spatial imaging resolution and 100 ns temporal resolution was used. The shadowgraphic layout consists of a CCD detector with a lens and nanosecond-pulsed backlight laser. An example of a shadow microphotograph is presented in Fig. 8.13.

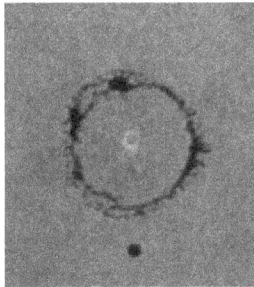

▲ Fig. 8.13.   Image of droplet fragment taken 2 μs after the droplet-pulsed irradiation. White spot corresponds to the optical emission of laser-induced plasma with the dark region, covered by initial droplet (at 0 ms delay).

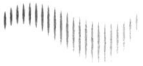

The regenerative amplifier system, Spitfire Pro (Spectra Physics, Inc.), which generated laser pulses with duration of 120 fs and central wavelength of 800 nm, was used in experiments described below as a source of optical radiation for excitation of the microdroplet target. The microdroplet was excited by two co-directional delayed laser pulses with the same linear horizontal polarization and different energies. Most of the experimental results presented below have been acquired under excitation of the microdroplet target with a 30 µJ energy pre-pulse, and 600 µJ energy main pulse delayed to 166 ps. During the experiments, the repetition rate of the laser system was decreased to 4 Hz to minimize pollution of the dispenser device with the scattered droplet fragments. To verify the absence of damping of registered THz signal due to the possible deposition of the scattered droplet debris on the wall of PP tube housing, we controlled its transmission before and after the laser-droplet interaction.

THz radiation was registered by a helium-cooled general purpose silicon bolometer (Infrared Labs, Inc.). The bolometer detector was positioned on a specially designed rotation stage with circular steel rail, which provided the capability of manually controlled rotation around the point of laser-droplet interaction (see Fig. 8.12) with an accuracy of combining the droplet position and the rotation axis around 0.5 mm. THz radiation was collected with the use of a PP lens, $F = 15$ cm, in the scheme 2F-2F. Spectral range of registered THz signal was limited to 3 THz due to the filters incorporated into the bolometric detector and transmission of PP lens and PP housing of the vacuum chamber. The THz radiation was collected in the angle range of 0°–150°, because of limitations set by three input/output ports placed at the PP housing tube, and the signal collection angle was defined by the focal length and aperture of PP lens as 20°.

X-ray radiation jointly generated with the THz signal was detected with the help of a photomultiplier tube PMT-119 with NaI scintillator. Energy of detected X-ray quanta is limited by transmission of the plastic wall of the vacuum chamber and starts from 7 keV at the transmittance level of 10^{-2}. X-ray detector was placed at the angle of 150° relative to optical beam direction. Spatial distribution of the THz signal was measured for different positions of the bolometric detector in the angle range of 0°–150° and under fixed placement of the X-ray detector.

Using the shadowgraph technique it was shown that the microdroplet keeps its form during tens or even a hundred nanoseconds after irradiation with the femtosecond laser pulse. Therefore, one should deal with the sub-wavelength THz radiation source, located near the sub-wavelength opaque metallic sphere. So diffraction of the emerged THz radiation on the metallic sphere should be taken into account. Also, there is a probability of THz emission by the transient currents on the whole surface of the microdroplet. Thus, a detailed investigation of angular distribution of the emitted THz radiation can shed some light on the origin of radiation from the microdroplets. Many times the measuring of angular distributions of THz signal generated from wire [24–27] and foil [28–30] targets were reported, and different excitation conditions lead to the gradual change of the THz energy distribution. For the jet source, maximum signal was detected in the direction of 50° [31] to the incident pump laser beam. The authors explained it as a result of interference of THz generation and absorption in the water jet. In our case the microdroplet target is even less transparent, which increases the role of THz diffraction.

8.3.1.2 THz Signal Dependence on Excitation Scheme and Laser Pulse Parameters

For the given laser and focusing parameters, single-pulse excitation did not provide any detectable THz signal. Hence, double-pulse excitation scheme was used (see Fig. 8.11). Fig. 8.14 represents the dependence of measured THz signal on time delay between two laser pulses. This curve was measured at zero angle of THz detection signal (see Fig. 8.11), i.e., in a forward direction relative to the laser pump beams. For zero delay between laser pulses and for single-pulse excitation, the THz signal was negligible. In the case of applying low-energy pre-pulse, the THz signal grew rapidly, then when the delay reached 20–50 ps the growing slowed down, and finally it came to saturation at the delay near 100 ps. The use of second harmonic pre-pulse did not change the form of the curve, and zero

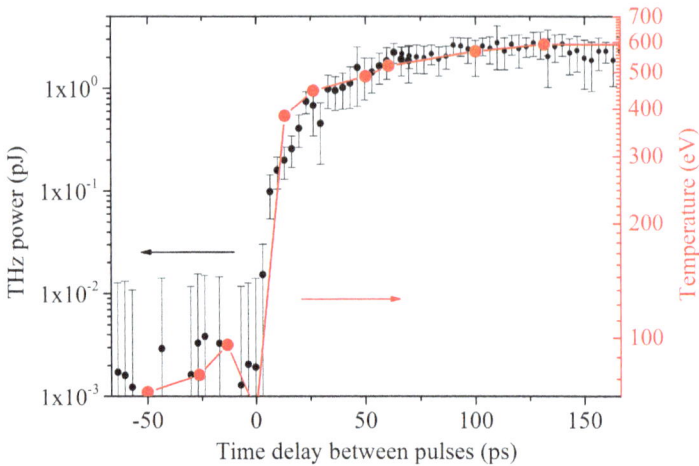

▲ Fig. 8.14.   Measured THz power (black) and calculated maximal temperature of electrons (red) depending on the delay between laser pulses for double-pulse regime of microdroplet excitation. Reprinted from Ref. [5] with permission.

delay also corresponded to the minimum in the detected signal. Results of numerical calculations on temperature of electrons on the surface of the droplet showed good correlation with tendency in behavior of dependence of THz signal registered in experiment (see Section 8.3.1.4 and Fig. 8.14).

Thus, by varying the time delay and energy ratio between two laser pulses one can achieve optimal parameters for optical-to-THz conversion (see Fig. 8.15). We found the optimal energy ratio to be around 0.05–0.1, and the optimal time delay to be around 170 ps (see Figs. 8.15 and 8.14, respectively).

Values of THz and X-ray signals measured as a function of the focusing lens L1 position along the optical axis of the system are presented in Fig. 8.16. As mentioned above, the THz radiation was registered in multiple directions relative to the laser beam propagation direction, while the X-ray signal was detected in the direction of 150° only. It is clearly seen that both THz and X-ray signals reach maximal values near the focal point. Assuming the beam waist size in the focal point as $r = 15$ µm (measured with the help of CCD camera for the weakened laser beam),

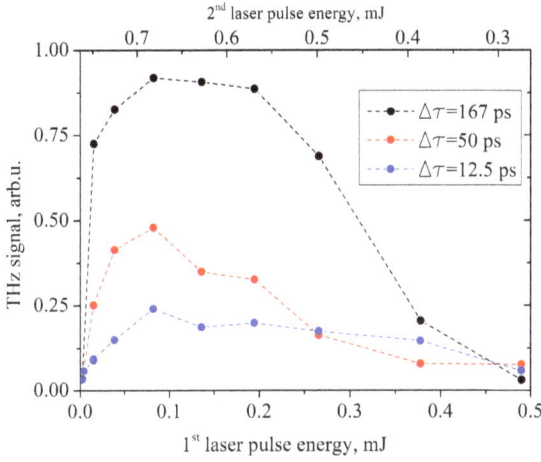

▲ Fig. 8.15. Dependence of THz signal on the ratio of laser pulse energies.

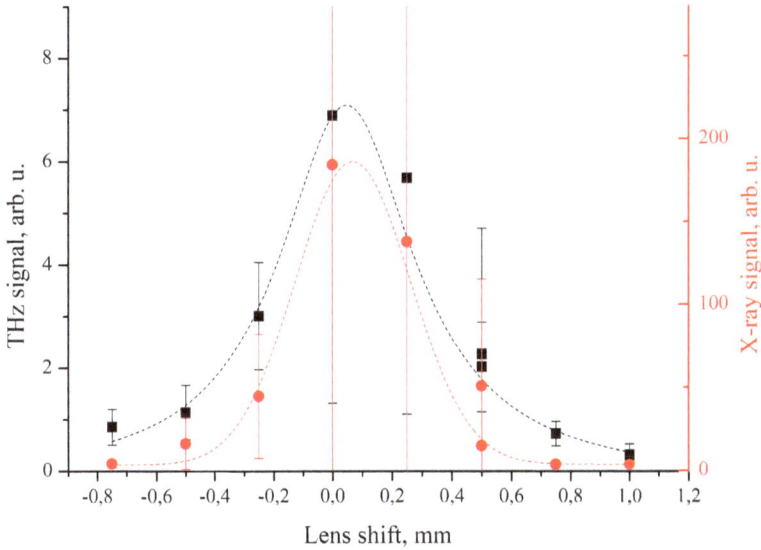

▲ Fig. 8.16. Dependence of THz (black dots) and X-ray (red dots) signals on shift of focusing lens L1. The lens was translated parallel to the optical axis of the system, and the droplet was located near the focal plane of the lens. Detection angle of THz radiation was 120° relative to the optical axis.

one can estimate the characteristic focal depth as $z_0 = \pi r^2/\lambda = 880$ mm. On the other hand, from Fig. 8.16 one can see that characteristic width of the THz and X-ray signal dependencies on the focusing lens position is nearly the same and its value is around 200 mm. The difference in the positions of the maximal values of the THz and X-ray signals is less than 100 mm, and position of the maxima corresponds to a case where the laser beam focus is located near the droplet's surface. The maximal size of droplet fragment cloud was also observed close to the zero shift position of the L1 lens.

It had been shown in [5] that one can manage the polarization characteristics of generated THz radiation by changing the polarization of the more powerful main optical pulse. The THz polarization was measured to be elliptical. Main axis of polarization ellipse for the THz radiation was co-directional with linear polarization of the main optical pulse. In Fig. 8.17 one can see the dependence of the total THz power and ellipticity of THz radiation on the direction of linear polarization of the main laser pulse. This polarization-dependency phenomenon could be explained in

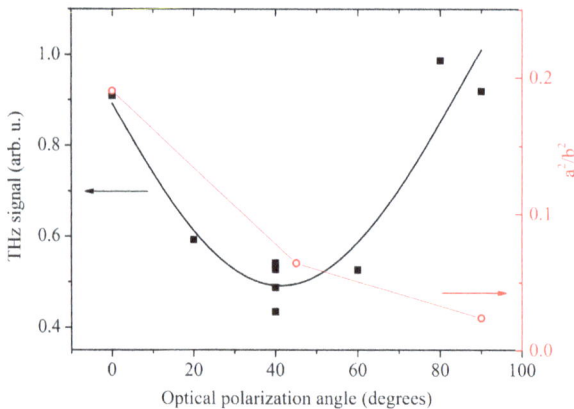

▲ Fig. 8.17.   Dependence of total THz power (black symbols) and ellipticity (square of the ratio of the polarization ellipse axes a and b) of THz radiation (red symbols) on polarization of main powerful laser pulse. Zero angle corresponds to horizontal direction of linearly polarized laser radiation. Adapted from Ref. [5] with permission.

the framework of nonlinear susceptibility of second order, assuming that the droplet's form is slightly different from the spherical one [5].

8.3.1.3 THz Radiation Pattern

It should be noted that values of both THz and X-ray signals registered in the experiments under the fixed excitation conditions varied significantly, while parameters of the droplet's excitation were similar and observed scenario of the droplet's fragmentation was also similar. Nevertheless, there is quite a good correlation between the THz and X-ray signals. Since conditions of the droplet excitation and parameters of the X-ray collection system remained fixed through the experiment, it was possible to sample the measured THz amplitudes into a few groups based on ranking of X-ray signal amplitudes registered jointly.

The THz radiation patterns were measured for the excitation of droplets of different size: "small" with diameter of 39 μm and "large" with diameter of 50 μm. The results of corresponding experiments are presented in Figs. 8.18 and 8.19: one can see a number of graphs sampled and grouped according to X-ray photon energies, such as 15–50, 200–300 and >400 arbitrary units. It is clearly seen that all these graphs exhibit a single maxima located close

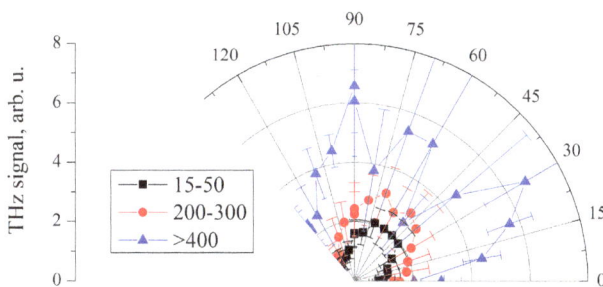

▲ Fig. 8.18.   Dependence of the THz signal from large droplet on the detection angle and energy of X-ray radiation. Droplet diameter $d_2 = 50$ μm. 15–50, 200–300 and >400 are ranges of energies of X-ray photons in arbitrary units.

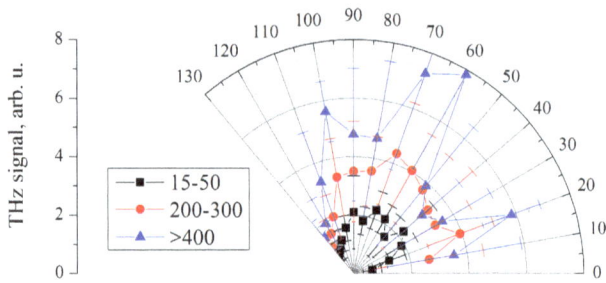

▲ Fig. 8.19. Dependence of the THz signal from small droplet on the detection angle and energy of X-ray radiation. Droplet diameter $d_1 = 39$ μm. 15–50, 200–300 and >400 are ranges of energies of X-ray photons in arbitrary units.

to 60° for low X-ray energies (15–50 and 200–300 arbitrary units), while the angular diagram of generated THz signal becomes more complex for higher energies of X-ray photons (>400 arbitrary units).

8.3.1.4 Numerical Calculations on Media Ionization and Electron Dynamics in the Ionized Near-Surface Region of the Droplet

To explain the observed behavior of the THz generation, we estimated parameters of laser-produced plasma. A one-temperature one-fluid model of a quasi-neutral plasma was used for numerical calculations. The local degree of ionization was determined from the stationary distribution obtained in the framework of collisional-radiative equilibrium using the THERMOS toolkit [39]. This approximation is correct for considering the plasma dynamics on picosecond timescales. However, directly during the main pulse, the degree of ionization could be lower than that obtained in the stationary model. The equation of state was calculated using the Thomas-Fermi model with semi-empirical corrections in the phase transition region using the FEOS code [40]. The energy transfer due to thermal radiation was described in the multigroup diffusion approximation [41]. Group emissivity and Rosseland range were calculated using the THERMOS toolkit. The energy transfer by electrons was described by means of the classical heat equation, without taking into account ballistic electrons, which are generated in a noticeable amount at laser intensities above 10^{16} W/cm². The thermal conductivity coefficient was calculated according to the Spitzer model with correction in the region of cold dense matter [42]. The absorption of the initial laser pulse was assumed to be instantaneous, and the distribution of the absorbed power and the absorption coefficient for the 2D case were calculated similarly to the 1D case [5]. The absorption of the main pulse was calculated using a hybrid model that combines geometric optics with the solution of the wave equation near the critical surface [43].

Results of hydrodynamic numerical calculations carried out on the droplet's temporal evolution are presented in Fig. 8.20. Sub-wavelength

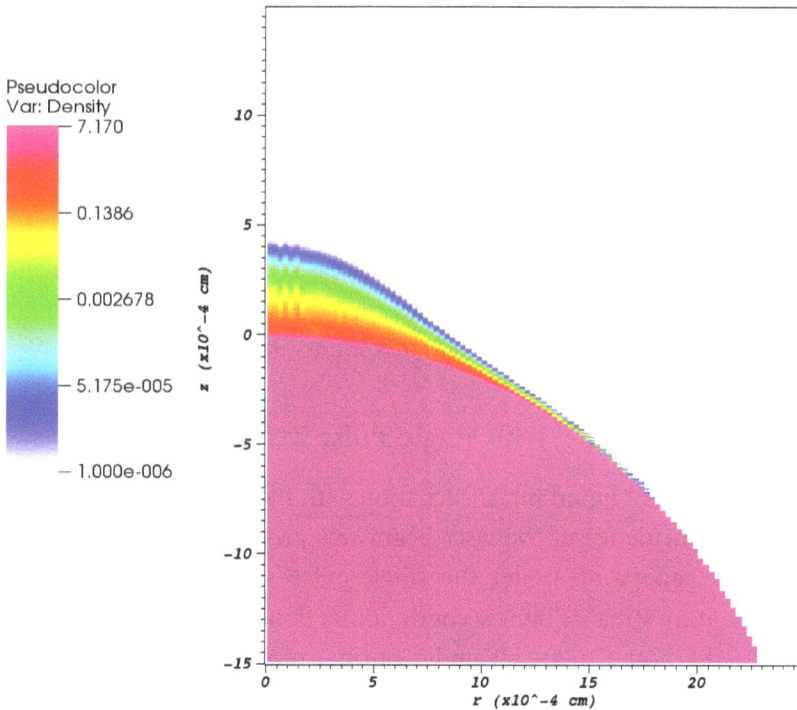

▲ Fig. 8.20.   Numerical calculation of droplet material density near the droplet's surface for 50 ps delay between pre- and main pulses.

droplet keeps its form during the few first nanoseconds of the laser-plasma interaction, so one can expect that the THz signal with wavelength of a few hundred microns will suffer diffraction on the sub-wavelength metallic sphere with a diameter of a few tens of microns. Precise calculation of the diffraction effect is quite complicated due to the sub-wavelength sizes of both the THz source and the distance from the source to the droplet. A proper calculation of the THz diffraction pattern is complicated and requires near-field computations, so here we provide only a qualitative model for its explanation.

Taking into account obtained experimental results, one may conclude that the most probable case is THz generation in near-surface ionized region with width less than $c\tau$, where τ is duration of laser pulse. Interference of contributions from the dipole and quadrupole currents [32] can provide a complicated structure with multi-lobe angular pattern, and the presence of an opaque thick droplet decreases the forwardly directed contribution into the THz signal. Laser-excited electrons in the plasma area have a complex energy distribution that leads to various THz and X-ray signals for identical laser excitation parameters.

Also, one can compare considered case with the ordinary optical rectification on metal surfaces [14]. In that case the THz wave propagates along the laser pump beam or the reflected beam, thus the signal should have a maximal value into the rear hemisphere.

8.3.2 Liquid Metal Jet with Single-Pulse Excitation

Another scheme of liquid metal excitation with the use of tighter focusing and grazing incidence of the laser beam was proposed by Cao *et al.* [6] to get THz generation under the single-pulse single-color interaction. A liquid gallium stream that was continuously flowing down due to gravity and formed a liquid metal jet under normal room conditions was used as the laser target in their experiments. It was shown that optical-to-THz conversion efficiency in liquid gallium stream could be even higher than in

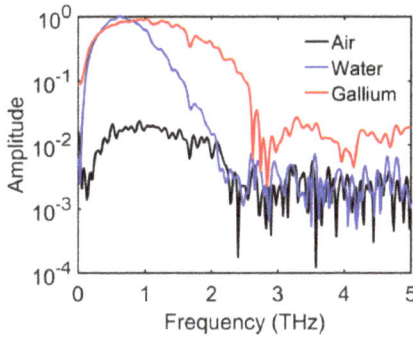

▲ Fig. 8.21.   THz generation spectra from liquid metal jet and other liquids. Reprinted from Ref. [4] with permission.

water stream target under the same excitation condition. Also the spectrum of generated THz radiation was found to be approximately two times wider in the first case (see Fig. 8.21).

It is necessary to note that laser excitation conditions are somewhat different for the liquid metal microdroplet *vs* the liquid metal jet. The laser radiation was directed normally to the surface of the droplet in [5], while the grazing incidence was key to acquiring considerable THz signal in [6], because the THz signal was quite tiny when the laser beam was directed perpendicular to the surface of the metallic jet. Because the thickness of the jet is around four times larger than the diameter of the microdroplet, it is rather complicated to provide a direct comparison of experimental results obtained for these two different targets.

8.3.3 Summary

The two different kinds of liquid metal targets have been studied to generate THz radiation under optical excitation. It has been shown that the use of liquid metal target can provide a wide generation spectrum and higher optical-to-THz conversion efficiency compared to air, for both microdroplet and jet "shapes" of the target. In addition, for the

microdroplet-based source, one can dynamically control the polarization and intensity of the emerged THz radiation. For future improvement and increasing the intensity of THz generation, it seems reasonable to survey different alloys of metals and to study different regimes for excitation of liquid metal targets with laser radiation.

Applying liquid metal as laser targets allows one to attain the joint generation of ultraviolet, X-ray, and THz radiation, which makes them promising for development of multi-frequency sources to provide coupled ultrashort electromagnetic pulses for extreme multi-spectral nonlinear laser applications.

8.4 Perspective on Investigation of New Liquid Media THz Sources

While gaseous THz photonics have been well studied, fluids are still "terra incognita" in the field of THz applications. With the first works on THz generation in liquids published in 2017 [44, 45] proving the possibility to create an effective source of THz radiation, theoretical and experimental study of fluids began. Due to a variety of features liquids possess and the considerable influence of the liquid medium on the laser pump pulses and the resulting THz radiation, one has to carefully choose the experimental methods for the formation of the fluid laser target and schemes for its pumping, taking into account that the properties of the liquid medium could be affected by the pump radiation as well. Therefore, it is necessary to investigate various classes of liquid media in order to find optimal conditions for the generation and extraction of THz radiation from the medium. In addition to classical liquids, it is also interesting to study supercritical fluids, since an increase in the nonlinear coefficients of such media has been shown [46]. The efforts expended on the search for new liquid media and methods for THz generation with use of sub-picosecond laser pump can pay off a hundredfold when they emerge as powerful, stable broadband sources of pulsed radiation in the THz spectral range.

References

1. Balakin A. V., *et al.* (2019). Terahertz wave generation from liquid nitrogen, Photonics Research, 7(6), pp. 678–686.
2. Balakin A. V., *et al.* (2018). "Terhune-like" transformation of the terahertz polarization ellipse "mutually induced" by three-wave joint propagation in liquid, Optics Letters, 43(18), pp. 4406–4409.
3. Cao Y., Ling F. & Zhang X.-C. (2020). Flowing cryogenic liquid target for terahertz wave generation, AIP Advances, 10(10), pp. 105119.
4. Zhang L., Tcypkin A., Kozlov S., Zhang C. & Zhang, X.-C. (2021). Broadband THz sources from gases to liquids, Ultrafast Science, 2021, pp. 9892763.
5. Solyankin P. M., *et al.* (2022). Single free-falling droplet of liquid metal as a source of directional terahertz radiation, Physical Review Applied, 14(3), pp. 034033.
6. Cao Y., *et al.* (2020). Broadband terahertz wave emission from liquid metal, Applied Physics Letters, 117(4), pp. 041107.
7. Francis K. G., *et al.* (2021). Forward terahertz wave generation from liquid gallium in the non-relativistic regime, Journal of the Optical Society of America B, 38(12), pp. 3639–3645.
8. Timmerhaus K. D. & Reed R. P. (Eds.). (2007). *Cryogenic Engineering: Fifty Years of Progress.* Springer.
9. Kildal H. & Brueck S. R. J. (1980). Orientational and electronic contributions to the third-order susceptibilities of cryogenic liquids, The Journal of Chemical Physics, 73(10), pp. 4951–4958.
10. Samios J., Mittag U. & Dorfmüller T. (1985). The far infrared absorption spectrum of liquid nitrogen: a molecular dynamics simulation study, Molecular Physics, 56(3), pp. 541–556.
11. Maker P. D., Terhune R. W. & Savage C. M. (1964). Intensity-dependent changes in the refractive index of liquids, Physical Review Letters, 12(18), pp. 507.
12. Gottfried B. S., Lee C. J. & Bell K. J. (1966). The Leidenfrost phenomenon: film boiling of liquid droplets on a flat plate, International Journal of Heat and Mass Transfer, 9(11), pp. 1167–1188.
13. Yiwen E., *et al.* (2018). Terahertz wave generation from liquid water films via laser-induced breakdown, Applied Physics Letters, 113(18), pp. 181103.

14. Kadlec F., Kužel P. & Coutaz J.-L. (2014). Optical rectification at metal surfaces, Optics Letters, 29(22), pp. 2674–2676.
15. Sagisaka A., *et al.* (2008). Simultaneous generation of a proton beam and terahertz radiation in high-intensity laser and thin-foil interaction, Applied Physics B, 90(3), pp. 373–377.
16. Jin Z., *et al.* (2016). Highly efficient terahertz radiation from a thin foil irradiated by a high-contrast laser pulse, Physical Review E, 94(3), pp. 033206.
17. Zhuo H. B., *et al.* (2017). Terahertz generation from laser-driven ultrafast current propagation along a wire target, Physical Review E, 95(1), pp. 013201.
18. Zeng Y., *et al.* (2020). Guiding and emission of milijoule single-cycle THz pulse from laser-driven wire-like targets, Optics Express, 28(10), pp. 15258–15267.
19. Nazarov M. M., *et al.* (2020). Measurements of THz and X-ray generation during metal foil ablation by TW, sub-relativistic laser pulses, Journal of Physics: Conference Series, 1692(1), pp. 012018.
20. Liao G., *et al.* (2019). Multimillijoule coherent terahertz bursts from picosecond laser-irradiated metal foils, Proceedings of the National Academy of Sciences, 116(10), pp. 3994–3999.
21. Krivokorytov M. S., *et al.* (2017). Cavitation and spallation in liquid metal droplets produced by subpicosecond pulsed laser radiation, Physical Review E, 95(3), pp. 031101.
22. Wagner C. & Harned N. (2010). Lithography gets extreme, Nature Photonics, 4(1), pp. 24–26.
23. Cunningham P. D., *et al.* (2011). Broadband terahertz characterization of the refractive index and absorption of some important polymeric and organic electro-optic materials, Journal of Applied Physics, 109(4), pp. 043505.
24. Deibel J. A., *et al.* (2006). Frequency-dependent radiation patterns emitted by THz plasmons on finite length cylindrical metal wires, Optics Express, 14(19), pp. 8772–8778.
25. Zhang D., *et al.* (2022). Towards high-repetition-rate intense terahertz source with metal wire-based plasma, IEEE Photonics Journal, 14(1), pp. 1–5.

26. Zeng Y., *et al.* (2020). Guiding and emission of milijoule single-cycle THz pulse from laser-driven wire-like targets, Optics Express, 28(10), pp. 15258–15267.
27. Zhuo H. B., *et al.* (2017). Terahertz generation from laser-driven ultra-fast current propagation along a wire target, Physical Review E, 95(1), pp. 013201.
28. Hu K. & Yi L. (2020). Relativistic terahertz radiation generated by direct-laser-accelerated electrons from laser-foil interactions, Physical Review A, 102(2), pp. 023530.
29. Liao G., *et al.* (2019). Multimillijoule coherent terahertz bursts from picosecond laser-irradiated metal foils, Proceedings of the National Academy of Sciences, 116(10), pp. 3994–3999.
30. Jin Z., *et al.* (2016). Highly efficient terahertz radiation from a thin foil irradiated by a high-contrast laser pulse, Physical Review E, 94(3), pp. 033206.
31. Zhang L.-L., *et al.* (2019). Strong terahertz radiation from a liquid-water line, Physical Review Applied, 12(1), pp. 014005.
32. Shkurinov A. P., *et al.* (2017). Impact of the dipole contribution on the terahertz emission of air-based plasma induced by tightly focused femtosecond laser pulses, Physical Review E, 95(4), pp. 043209.
33. Nazarov M. M., *et al.* (2008). Terahertz time-domain spectroscopy of biological tissues, Quantum Electronics, 38(7), pp. 647.
34. Boyd R. W. (2020). *Nonlinear Optics.* Academic Press.
35. Landau L. D. & Lifshitz E. M. (2013). *Course of Theoretical Physics.* Elsevier.
36. Sun W.-F., *et al.* (2008). External electric field control of THz pulse generation in ambient air, Optics Express, 16(21), pp. 16573–16580.
37. Kim K.-Y., *et al.* (2007). Terahertz emission from ultrafast ionizing air in symmetry-broken laser fields, Optics Express, 15(8), pp. 4577–4584.
38. Balakin A. V., *et al.* (2010). Terahertz emission from a femtosecond laser focus in a two-color scheme, Journal of the Optical Society of America B, 27(1), pp. 16–26.
39. Vichev I. Y., *et al.* (2019). On certain aspects of the THERMOS toolkit for modeling experiments, High Energy Density Physics, 33, pp. 100713.

40. Faik S., Tauschwitz A. & Iosilevskiy I. (2018). The equation of state package FEOS for high energy density matter, Computer Physics Communications, 227, pp. 117–125.
41. Zeldovich, I. B. & Raizer Y. P. (1966). *Physics of Shock Waves and High-Temperature Hydrodynamic Phenomena.* Academic Press.
42. Spitzer L. (2006). *Physics of Fully Ionized Gases.* Courier Corporation.
43. Basko M. M. & Tsygvintsev I. P. (2017). A hybrid model of laser energy deposition for multi-dimensional simulations of plasmas and metals, Computer Physics Communications, 214, pp. 59–70.
44. Jin Q., *et al.* (2017). Observation of broadband terahertz wave generation from liquid water, Applied Physics Letters, 111(7), pp. 071103.
45. Dey I., *et al.* (2017). Highly efficient broadband terahertz generation from ultrashort laser filamentation in liquids, Nature Communications, 8(1), pp. 1–7.
46. Mareev E., *et al.* (2018). Anomalous behavior of nonlinear refractive indexes of CO_2 and Xe in supercritical states, Optics Express, 26(10), pp. 13229–13238.

https://doi.org/10.1142/9789811265648_0009

Chapter 9
Ultrafast Dynamics in Liquids Excited by Intense Terahertz Waves

Liangliang Zhang, Minghao Zhang, Wen Xiao, Hang Zhao, Cunlin Zhang

Key Laboratory of Terahertz Optoelectronics (MoE), Department of Physics, Capital Normal University, China

9.1 Introduction

As the most important carrier in the life sciences, liquid water can combine with various solutes to form solutions, resulting in more complex molecular motions. A complex aqueous solution, containing various substances, provides conditions for biochemical reactions and is closely related to important fields such as biomedicine and the ecological environment. Low-frequency molecular motion associated with hydrogen bond networks in water/aqueous solutions plays a crucial role in these biochemical reactions. Understanding the existing forms [1–9], properties, and behaviors [10–12] of the water network structure at the molecular and atomic levels is of great significance for confronting the fundamental issues in areas related to liquid water. Therefore, since the discovery of hydrogen bonds in the last century, researchers in various fields, such as the natural sciences, have explored molecular dynamics in water through various theoretical and experimental techniques.

Polarizability $\chi(\omega)$ reflects the collective/cooperative molecular motions at frequencies in the gigahertz (GHz) to terahertz (THz) range

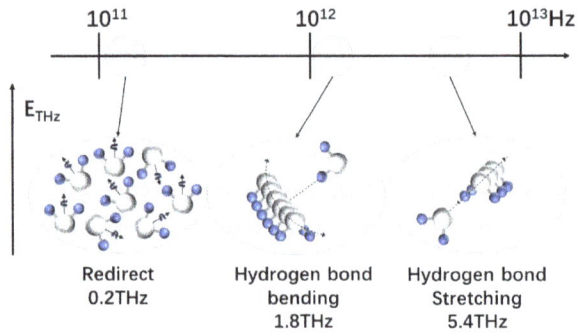

▲ Fig. 9.1. Dominant modes of molecular motion of water in the THz frequency range.

and is strongly influenced by intermolecular mode dynamics. Therefore, evaluating the dielectric sensitivity of liquid water using optical methods, especially nonlinear THz techniques, has always been an important means to explore its hydrogen bonding-related dynamic properties. The THz Kerr effect (TKE) technology breaks through the bottleneck of traditional spectroscopic methods [13–23], providing new ideas for exploring low-frequency molecular motion. This technique relies on an incident THz electric field resonating with the rotational transition of a single molecule or the collective/coordinated low-frequency motion of the molecule. The low-frequency molecular motion is highlighted, and the time-resolved evolution of liquid anisotropy is directly observed. The motion associated with hydrogen bonds in liquid water is extremely sensitive to THz radiation (Fig. 9.1). TKE is thus expected to achieve ultrafast time-resolved kinetic observations of the hydrogen bond network in liquid water, which also lays the foundation for exploring the interaction between hydrogen bonds and solutes in aqueous solutions.

9.2 Kerr Effect of Liquid Water Driven by THz Field

A gravity-driven flowing liquid film was applied in the TKE measurements. A schematic of the experimental setup is shown in Fig. 9.2. A Ti:sapphire laser amplifier (Spitfire-Ace) was used as the laser source. The laser pulse

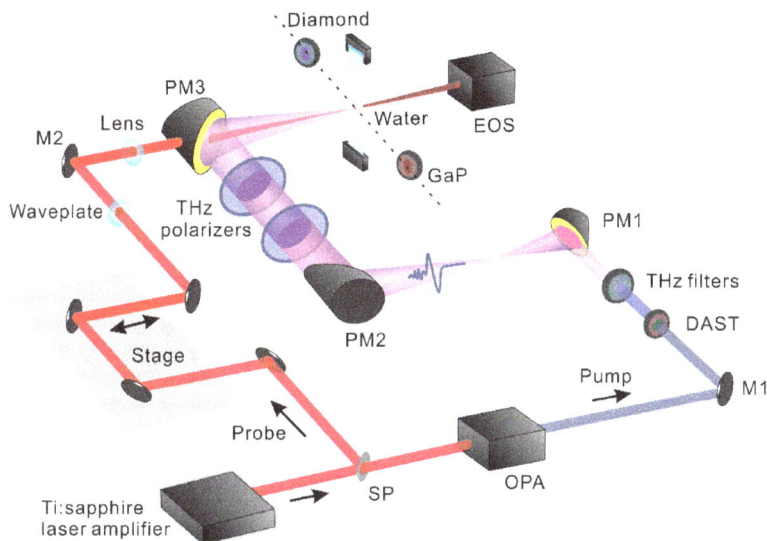

▲ Fig. 9.2.   Schematic diagram of the TKE measurement system (the water film can be replaced by diamond or gallium phosphide crystal). SP: beam splitter; OPA: optical parametric amplifier; M: plane mirror; PM: off-axis parabolic mirror; EOS: electro-optical sampling; DAST: 4-N,N-dimethylamino-4'-N'-methyl-stilbazoliumtosylate.

passed through an optical parametric amplifier (OPA, Spectra Physics) to generate a laser pulse with a center wavelength of 1550 nm and a repetition frequency of 1 kHz. The femtosecond pulse excites an organic 4-N,N-dimethylamino-4'-N'-methyl-stilbazoliumtosylate (DAST) crystal to generate strong THz waves. The residual near-infrared pulse in the optical path was filtered out by a set of low-pass filters with a cut-off frequency of 18 THz. The THz pulse energy is controlled via two THz polarizers and then reflected and focused onto the sample surface by an off-axis parabolic mirror (PM3). The probe pulse polarized at 45° relative to the vertically polarized THz pulse was co-focused on the sample with the THz pulse, thereby generating electro-optic modulation. Subsequently, a balanced photodiode was used after a quarter-wave plate and Wollaston prism to form an electro-optical sampling (EOS) differential analyzer to detect the birefringence signal induced by the THz electric field. The time trace was recorded by scanning the delay between the pump and the probe beams.

The THz time-domain waveform was obtained by EOS of a 100 μm thick gallium phosphide (GaP) crystal. Since the strong THz pulse generated by the DAST crystal may saturate the response signal of the GaP crystal in the experiment, the birefringence signal of a diamond crystal was used to calibrate the THz electric field strength [24]. In addition, in the electro-optical detection crystal, the response curve for a broadband THz pulse is distorted by factors such as frequency-dependent phase mismatch, reflection, dispersive propagation, and absorption. Therefore, the fully complex response function of the GaP detector was used in this study to reconstruct the THz spectrum [25–27]. The obtained THz pulse had a peak electric field strength of 14.9 MV/cm, spectral width of 1–10 THz, and center frequency of 3.9 THz.

9.2.1 TKE Responses of Water Excited by Different THz Electric Field Strengths

Fig. 9.3(a) shows the TKE response of water excited by different THz electric field strengths. The energy of the THz wave was controlled by

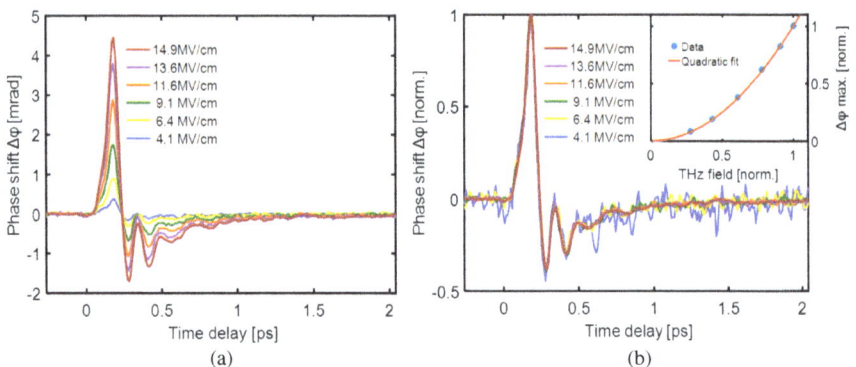

▲ Fig. 9.3. (a) TKE responses of water film excited by different THz electric fields and (b) normalized TKE responses; the inset shows the relationship between the THz electric field and the peak signal. (a) and inset of (b) are reprinted from Ref. [34] with permission.

adjusting two THz polarizers. The bipolar TKE signal of water with distinct oscillatory characteristics was experimentally observed with a fast-rising peak and a slow recovery time of ~1 ps. The signals in Fig. 9.3(b) show the normalized results. The characteristic curves displayed were almost coincident, indicating that no more nonlinear modes were introduced in the variation range of the THz energy. The inset shows the response peak under different THz wave excitations, which is proportional to the square of the THz electric field. This reveals that the Kerr effect dominates this process. The measured TKE signal is bipolar and accompanied by a distinct oscillatory signature that extends to ~1 ps.

Considering that the THz frequency band in this work mainly covers two modes of intermolecular hydrogen bond motion, namely, hydrogen bond bending and stretching vibration, the measured transient birefringence signal is mainly attributable to the water intermolecular structural dynamics. The phase shift $\Delta\varphi$ is positively correlated with the birefringence Δn; that is, $\Delta\varphi(t) = 2\pi/\lambda_{pr}\int_0^l \Delta n(z,t)dz$, where λ_{pr} is the probe beam wavelength. The birefringence response can be divided into electron and molecular contributions. The electron contribution can be expressed as the product of the square of the electric field multiplied by the wavelength and response factor B_e; that is, $\Delta n_e = \lambda B_e |E(z,t)|^2$. The electron contribution in water is usually considered a small response [28–30]. Regarding the molecular contribution, the relaxation and hydrogen bond vibrational processes affect the birefringence signal. In other words, $\Delta n_m = \Delta n_D + \Delta n_i$ ($i = 1$ and $i = 2$ represent hydrogen bond bending and stretching vibrations, respectively), where Δn_D is the birefringence caused by the Debye relaxation process.

The molecular dynamics mechanism has been well explained in previous studies [28, 31, 32]. However, the Debye relaxation process is weak when excited by THz waves with high-frequency components. Therefore, the TKE response is primarily derived from the movement of intermolecular hydrogen bonds. Comparative measurements of diamond and water were performed to further confirm that the birefringence signal of liquid water is mainly due to hydrogen bond motion rather than

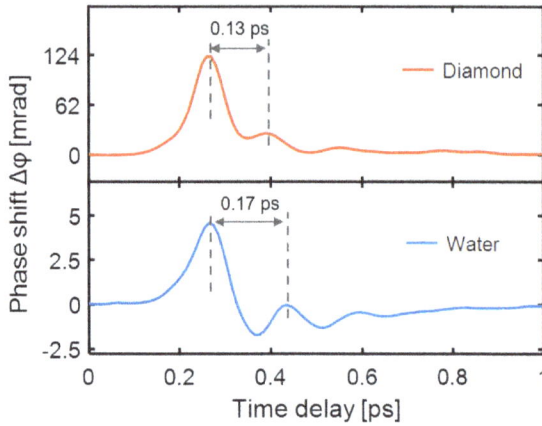

▲ Fig. 9.4. TKE responses of diamond and water. Reprinted from Ref. [34] with permission.

electron contribution. As shown in Fig. 9.4, the TKE response of diamond originates from the electronic contribution, which has a positive polarity and exhibits a double-peak structure with a time interval of ~0.13 ps. In contrast, the TKE signal of liquid water has a distinct negative polarity and a broadened double-peak interval of ~0.17 ps.

In theory, the TKE response of diamond completely follows the square curve of the THz electric field [24, 33]. The time interval of the double peaks ($\frac{1}{2}f = 0.13 ps$) with typical electronic response characteristics corresponded well with the center frequency of the THz pulse (f = 3.9 THz). The electronic contribution to the refractive index anisotropy depends entirely on the strength of the THz electric field, that is, $\Delta n_e = \lambda B_e |E(z,t)|^2$. The relationship between the phase shift of the probe light at 800 nm and the refractive index anisotropy was established by the following equation: $\Delta \varphi_e(t) = \frac{2\pi}{\lambda} \int_0^l \Delta n_e(z,t)dz$, where λ is the wavelength of the probe light, and l is the thickness of the sample. In addition, in the TKE experiment, some factors affect the electronic response, such as the sampling error, phase shift, and attenuation of the THz wave during transmission in the sample [24, 25, 34, 35]. Considering the above influencing factors, the electron response equation can be extended as

$$\Delta\varphi_e(t) = 2\pi B_e \int_0^l \left[t_{12} E(t+\beta x)\cdot\exp(-\alpha x)\cdot\exp(i\arctan(n_{THz}x/Z_R)) \right]^2$$

$$\times dx * \exp(-4\ln 2(t/\tau)^2). \tag{9.1}$$

B_e is the electron response factor, which is approximately 0.00996×10^{-14} m/V^2 for diamond [24, 28]. $t_{12} = 2/(n_{THz} + 1)$ stands for Fresnel transmission coefficient. n_{THz} represents the average refractive index of the THz waves, which is approximately 2.38, and β is the phase mismatch factor, which is approximately 0.04 ps/mm. α represents the average absorption coefficient of the THz waves, which can be ignored for diamond. $e^{i\,arctan}(n_{THz}x/Z_R)$ represents the phase shift term. Z_R represents the Rayleigh length, which is about 0.26 mm in this work. $e^{-4\ln 2(t/\tau)^2}$ denotes the sampling pulse, and τ is the pulse width of the probe light, which is approximately 50 fs.

The fitted electronic responses for diamond and liquid water based on the method described above are shown in Fig. 9.5. The simulated electronic response in diamond agrees with the experimental data. However, owing to the small electronic response factor and short effective medium length, the electronic response of water is much weaker than that of diamond. Indeed, the electron response factor of a diamond crystal is

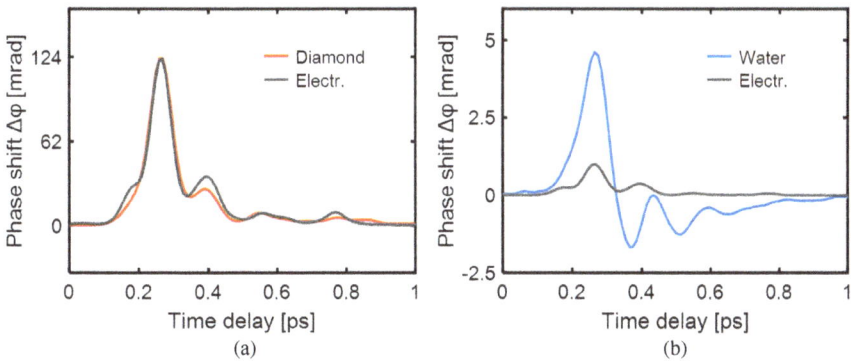

Fig. 9.5. TKE responses of (a) diamond and (b) water. The grey line represents the simulated electronic response. (a) is reprinted from Ref. [34] with permission.

more than 3.3 times that of liquid water ($< 0.003 \times 10^{-14}$ m/V² in water [28]). Water has a large absorption coefficient in the THz band, and water with a thickness of approximately 25 μm can absorb 90% of THz energy. Considering the thicknesses of both samples, the intensity of the electron response in a diamond with a thickness of 300 μm was estimated to be two orders of magnitude higher than that in liquid water with a thickness of 90 μm. The maximum phase shift induced by the electronic response was approximately 122 mrad in diamond and about 1.2 mrad in water, which is consistent with the estimation.

9.2.2 Frequency Dependence of the TKE Response

The TKE response can be resolved into electronic and molecular responses caused by the collective movement of molecules and intermolecular hydrogen bonds. Fig. 9.6(a) shows the simulation of the response induced by molecular motion (red line), intermolecular hydrogen bond stretching motion (purple line), intermolecular hydrogen bond bending motion (yellow line), and electronic relaxation (blue line). Fig. 9.6(b) shows the sum of the molecular and electronic responses. The overall simulation

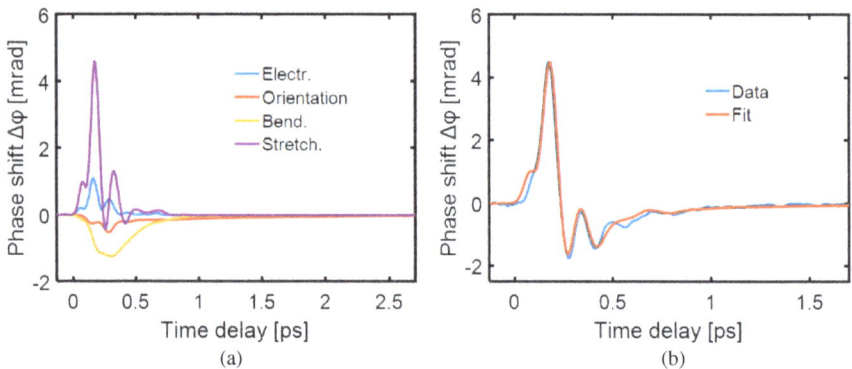

▲ Fig. 9.6.   (a) The simulated electron response (blue) and molecular directional motion (red) of liquid water, as well as the molecular response induced by intermolecular hydrogen bond stretching (purple) and bending (yellow) vibrations. (b) Measured TKE response (blue) and theoretical fit (red). Reprinted from Ref. [34] with permission.

results agree with the measured data. The values of the damping coefficient and inherent frequency used to obtain the best fit during the simulation were $\gamma_1 = 115$ cm^{-1}, $\gamma_2 = 165$ cm^{-1}, $\omega_1 = 60$ cm^{-1}, and $\omega_2 = 190$ cm^{-1}. This is consistent with the previous results obtained by Raman spectroscopy [21, 22]. Here, parameter a_i is defined to represent the relationship between the THz electric field strength and the induced refractive index anisotropy in two hydrogen bond motion modes:

$$a_i = \frac{\beta_i \mu_0^2}{6\sqrt{2}k_B T n_0} \left(\frac{\partial \alpha_{//}}{\partial q_i} - \frac{\partial \alpha_\perp}{\partial q_i} \right). \tag{9.2}$$

According to the experimental data, a_1 and a_2 can be calculated as 5.6×10^3 m^{2}s^2/V^2 and 5.2×10^4 m^{2}s^2/V^2, respectively.

In addition, different low-pass filters with cut-off frequencies of 9 and 6 THz were used to change the bandwidth of the driving THz pulse. The experimental measurements and simulation results are presented in Fig. 9.7. The molecular contributions exhibit a strong frequency dependence due to the involvement of different molecular motion mechanisms. The induced transient anisotropy mainly originates from the stretching and bending vibrations of intermolecular hydrogen bonds. As the high-frequency components were confined, the contributions of low-frequency molecular motions were significantly enhanced. The proportion of negative responses in the measurements gradually increased.

Under the excitation of a high-frequency THz pulse (e.g., central frequency ~3.9 THz), the contribution of the water molecular orientation process to the anisotropy is insignificant, and the intermolecular motion modes with apparent damping properties dominate. It can be predicted that the contribution of motion (such as the translation of water molecules) to the birefringent response needs to be considered when excitation with higher-frequency THz pulses (e.g., center frequency of ~11.4 THz) is used [22, 29]. The TKE responses obtained by varying the THz electric field frequency components were extracted

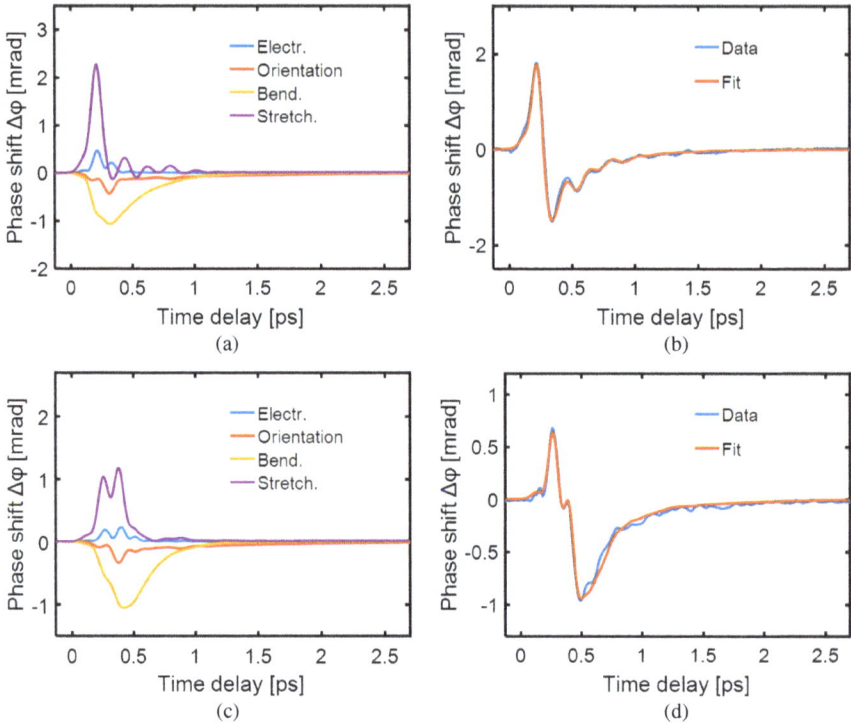

▲ Fig. 9.7.   The simulated electronic responses and molecular directional motions, as well as the molecular responses induced by intermolecular hydrogen bond stretching and bending vibrations of liquid water excited by THz pulses with a bandwidth of (a) 9 THz and (c) 6 THz. (b) and (d) Measured TKE response (blue line) and theoretical results (red line) corresponding to (a) and (c), respectively. Reprinted from Ref. [34] with permission.

using the Kerr coefficients in the 0.5–10 THz frequency domain, as shown in Fig. 9.8(a).

The Kerr response coefficients in this frequency range are mainly provided by the intermolecular modes. A significant resonance peak appears around 4.5 THz, which corresponds to the hydrogen bond stretching vibrational mode. The hydrogen bond bending vibration is approximately a critically damped harmonic oscillator; therefore, no

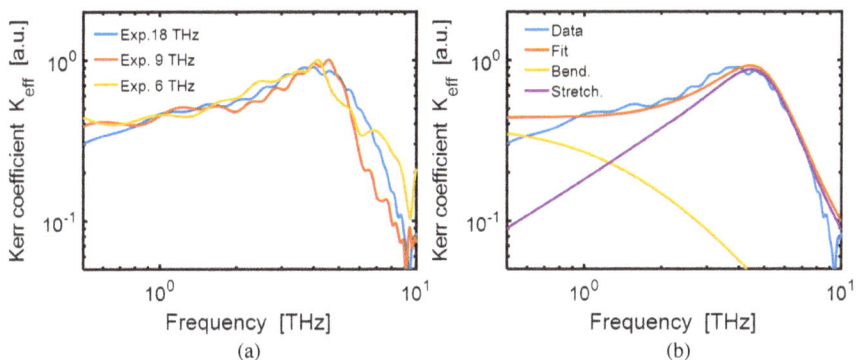

▲ Fig. 9.8. (a) Experimental measurement of the Kerr coefficient in the THz band. (b) Theoretical simulation by analyzing the bending and stretching vibrational modes of intermolecular hydrogen bonds.

prominent peak position is observed. The experimentally measured Kerr coefficients were compared with the theoretically predicted results. As shown in Fig. 9.8(b), the frequency response in the simulated intermolecular mode was mainly derived from the homogeneous solution of Eq. (9.3). The characteristic parameters are applied with the best fitting parameters in the time-domain simulation process, and the frequency-domain expression of the Kerr coefficient is obtained as follows:

$$K_{eff}(\omega) = \sum_{i=1,2} \frac{|k_i|}{\omega_i^2 - \omega^2 + i\gamma_i\omega}, \qquad (9.3)$$

where the coupling coefficient is expressed by $k_i = \frac{a_i}{\lambda_{pr}}$, and its value is $k_1 \approx -7 \times 10^9$ C/$(V \cdot kg \cdot m)$ and $k_2 \approx 6.5 \cdot 10^{10}$ C/$(V \cdot kg \cdot m)$. The final simulation results agree with the experimentally measured data.

9.2.3 TKE Response of Heavy Water

To verify the theoretical model further, the TKE response of heavy water (D_2O) was measured and compared with that of liquid water (Fig. 9.9).

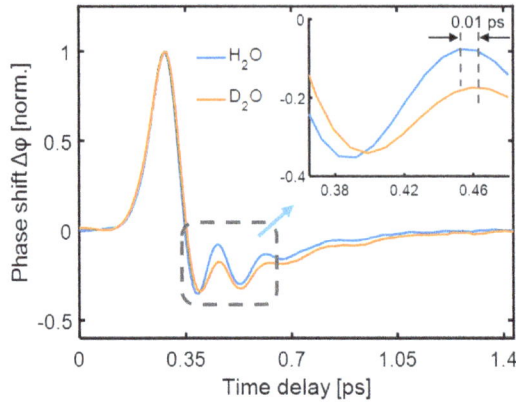

▲ Fig. 9.9. Comparison of TKE responses of water (H_2O) and heavy water (D_2O). Reprinted from Ref. [34] with permission.

The characteristic frequencies of the hydrogen bond stretching vibrations in heavy water were slightly redshifted compared to those in water, resulting in a slight broadening of the double-peak time interval of the TKE response. Furthermore, the relatively large damping coefficient of heavy water in the stretching mode corresponds to a faster energy decay process of the harmonic oscillator, slightly decreasing the second peak of the TKE response relative to water.

In addition, the TKE response of heavy water was measured by driving the THz pulses at different frequencies, as shown in Fig. 9.10(a). When the high-frequency components are limited, the proportion of negative responses in the signal gradually increases. The TKE responses of heavy water were assigned to various motion modes using the theoretical model proposed above, as shown in Figs. 9.10(b)–(d). The fitting results (shown in the inset) agreed well with the experimental measurements. In analyzing the molecular dynamics of heavy water, the parameters used to obtain the fitting curve were $\gamma_1 = 130$ cm^{-1}, $\gamma_2 = 175$ cm^{-1}, $\omega_1 = 60$ cm^{-1}, and $\omega_2 = 180$ cm^{-1}.

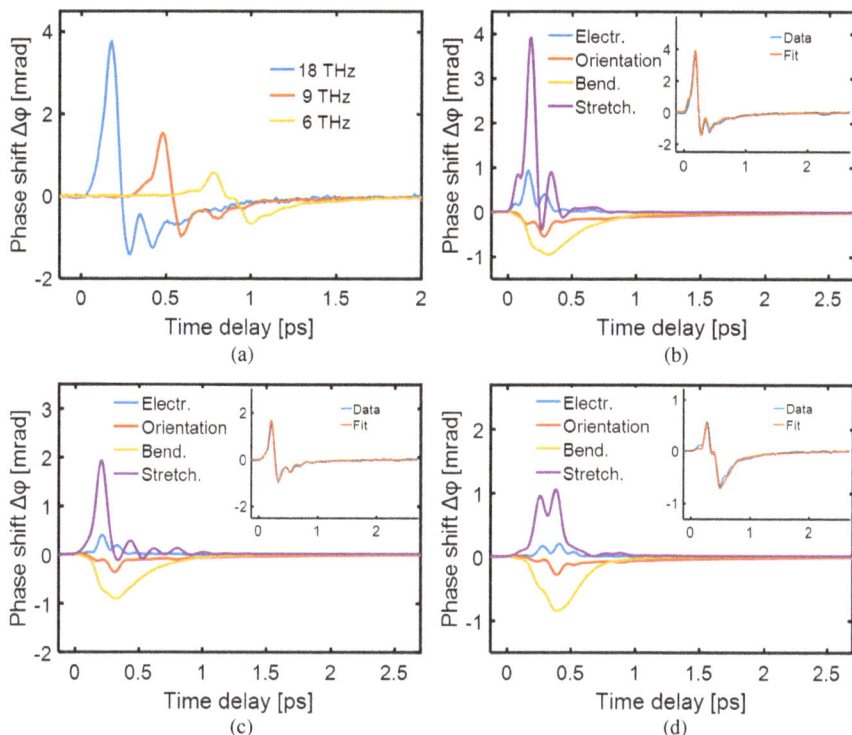

▲ Fig. 9.10. (a) TKE responses of heavy water excited by THz pulses passing through different low-pass filters. Time-resolved evolution of various low-frequency molecular motions in liquid water excited by THz pulses with a bandwidth of (b) 18 THz, (c) 9 THz, and (d) 6 THz. The insets show results of the theoretical fitting. (b)–(d) are reprinted from Ref. [34] with permission.

9.3 Kerr Effect of Ionic Aqueous Solution Driven by THz Field

Low-frequency molecular motion in aqueous solutions has significant implications in biological activity, affecting many processes, such as solvation, energy transfer, and proton transport. However, various anomalous properties resulting from molecular motions are not fully understood. For example, ionic and liquid water have different thermodynamic properties, including viscosity, and melting and boiling points. Although numerous theoretical and experimental studies have been conducted on the physical properties of liquid water, the microscopic mechanism of the water hydrogen bond network under ionic perturbation has not been fully elucidated.

This section uses the TKE technique to assess the low-frequency molecular motion of water under ionic perturbations. We used strong-field broadband THz pulses to excite various ionic aqueous solutions and observed the effect of ions on the TKE response of water. First, the TKE responses of aqueous solutions of sodium ions (Na^+) with different anions were investigated. Aqueous solutions of chloride ions (Cl^-) with various cations were then systematically measured by TKE. Na^+ and Cl^- were used as representative ions because the interaction between these ions and water molecules is equivalent to that between adjacent water molecules [30]. The research presented in this section provides a new perspective with regard to the ionic effect on the hydrogen bond network of water.

9.3.1 Sodium Chloride Solution

An intense broadband THz pulse (center frequency of 3.9 THz, peak electric field strength of 13.8 MV/cm, and bandwidth of 1–10 THz) was used to excite a free-flowing liquid film with a thickness of 90 μm, as shown in Fig. 9.11. A time-resolved phase shift $\Delta\varphi(t)$, which is proportional to the

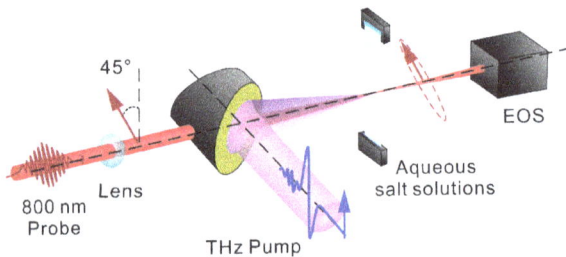

▲ Fig. 9.11. Schematic diagram of the TKE system for measuring aqueous salt solutions. Reprinted from Ref. [34] with permission.

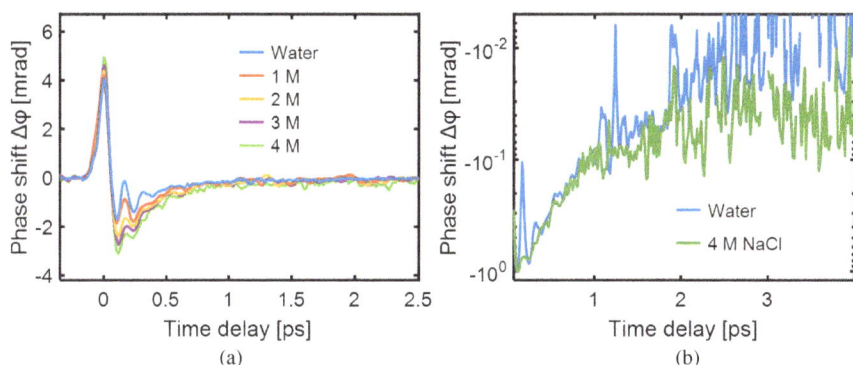

▲ Fig. 9.12. (a) TKE responses of water and NaCl solutions with different concentrations. (b) An enlarged logarithmic view of the negative polarity responses. Reprinted from Ref. [36] with permission.

transient birefringence $\Delta n(t)$ induced by the THz pulse, was detected using 800 nm femtosecond laser pulses.

The TKE responses of sodium chloride (NaCl) aqueous solutions with concentrations of 1, 2, 3, and 4 M (mol/L) were measured. The results are shown in Fig. 9.12(a). As the ion concentration increased, the characteristics of the TKE response exhibited three evident changes: (i) the response amplitude increased, (ii) the relaxation time increased, and (iii) the oscillation characteristics gradually smoothed. We normalized the maximum values of the negative polarity responses of water and the 4 M NaCl solution, and found that there was a slight difference in the responses after 1 ps, as shown in Fig. 9.12(b), proving that ions slow the recovery time of the molecular motions of water. To clarify the origin of this phenomenon, the TKE signal was assigned to different molecular motion patterns.

The TKE signal can be decomposed into weak electronic and dominant molecular responses. The electronic response depends on the hyperpolarizability of the water molecule, and the effect of ions on the electronic response in water is negligible [28, 31]. Here, we focus on the molecular response of the ionic aqueous solution, which can be assigned to Debye relaxation related to the molecular reorientation motion, intermolecular hydrogen bond bending, stretching modes associated with restricted translational motion, and a specific ion-water hydrogen bond vibration mode.

(1) Debye relaxation process. The reorientation motion of water molecules contributes to the negative polarity component of the TKE signal, which can evolve over several picoseconds [28]. As demonstrated experimentally, the addition of ions slows the reorientation dynamics of water molecules [32, 33]. A rotational-diffusion model [28] was used to fit the relaxation tail of the Debye relaxation response. As shown in Fig. 9.13(a), the obtained relaxation time constant increased almost linearly with an

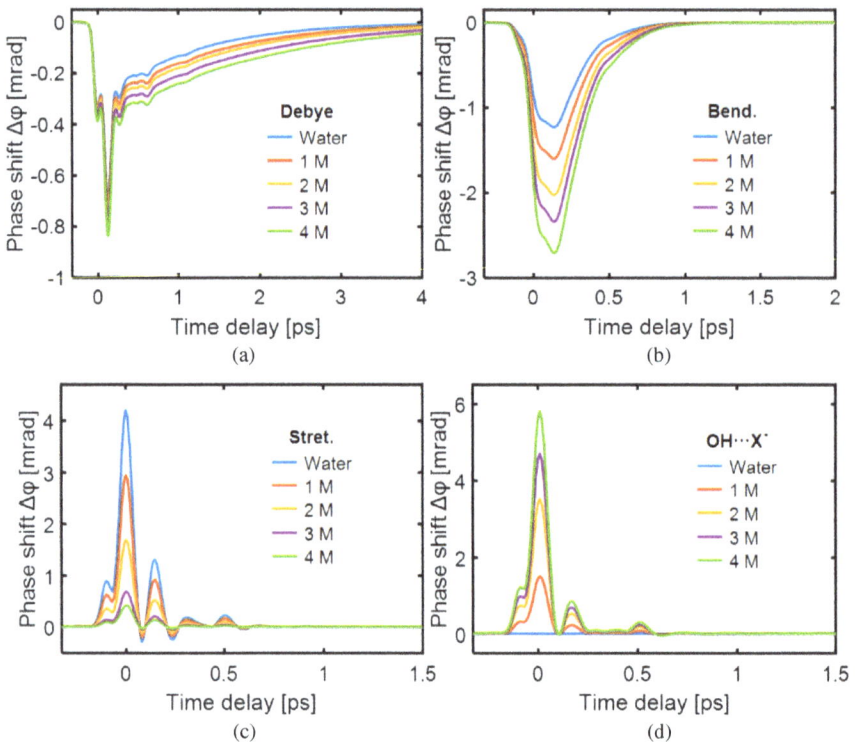

▲ Fig. 9.13. The simulated molecular responses of (a) Debye relaxation, (b) hydrogen bond bending, (c) hydrogen bond stretching, and (d) OH⋯Cl⁻ hydrogen bond vibration for water and NaCl solutions with different concentrations. Reprinted from Ref. [36] with permission.

increase in the solution concentration. The fitted values of the relaxation time constants are ~1.1 ps for water, ~1.28 ps, ~1.45 ps, ~1.62 ps, and ~1.8 ps for 1 M, 2 M, 3 M, and 4 M NaCl solutions, respectively.

(2) Intermolecular hydrogen bonding modes. In pure water, the THz electric field is coupled into restricted translational motions perpendicular and parallel to the hydrogen bond direction [34], resulting in a bipolar response. Here, the hydrogen bond harmonic oscillator model was used to simulate the anisotropic response under the intermolecular hydrogen bond bending and stretching modes [34, 35].

$$\frac{\partial^2 q_i(t)}{\partial t^2} + \gamma_i \frac{\partial q_i(t)}{\partial t} + \omega_i^2 q_i(t) = a_i E^2(t) \quad (i = 1, 2), \tag{9.4}$$

where γ_i represents the damping coefficient with values of $\gamma_1 = 115$ cm^{-1} for the bending mode and $\gamma_2 = 165$ cm^{-1} for the stretching mode. ω_i denotes the inherent frequency with values of $\omega_1 = 60$ cm^{-1} for the bending mode and $\omega_2 = 190$ cm^{-1} for the stretching mode. a_i is the coupling factor between the THz pulse intensity ($E^2(t)$) and driving force of the harmonic oscillator. q_1 and q_2 represent the anisotropic perturbations induced by the THz electric field under the bending and stretching modes, respectively, which satisfy the Lorentz dynamic equation associated with the damped harmonic oscillator motion. q_1 and q_2 are directly related to the refractive index anisotropy, that is, $\Delta n \approx b_2 q_2 - b_1 q_1$, where b_1 and b_2 represent the coefficients related to the dielectric susceptibility under the bending and stretching modes, respectively. For pure water, the positive polarity TKE response primarily originates from the stretching mode of intermolecular hydrogen bond motions. The negative polarity TKE response is due to the synergetic contributions of molecular reorientation and intermolecular hydrogen bond bending motion.

For an ionic aqueous solution, the migration of ions could provide an additional force field that affects the translational motions of water molecules perpendicular/parallel to the hydrogen bond direction under the applied THz electric field. For the simulated bending mode, as shown in Fig. 9.13(b), the amplitude of the negative polarity response gradually increases as the concentration increases. However, the relevant characteristic parameters, such as the damping coefficient and inherent frequency, did not change significantly in this over-damped oscillation system. For example, the simulated damping coefficient for the disturbed water-water hydrogen bond bending mode is 125 cm^{-1} for a 4 M NaCl solution, which is only slightly larger than that of pure water (115 cm^{-1}). The corresponding inherent frequency of 60 cm^{-1} is consistent with that of water, indicating that the addition of sodium and chloride ions is not sensitive to the hydrogen bond bending band of water, which is in agreement with a previous Raman study [37]. For the simulated stretching mode, which is shown in Fig. 9.13(c), the positive polarity response gradually decreases, along with the weakening of oscillation characteristics, as the concentration increases because the addition of ions dilutes the hydrogen bond density and destroys the water network structure. However, as the concentration increased, the positive polarity response increased, indicating that the interaction of ions with water provided a larger positive polarity response than that of the water-water hydrogen bond stretching mode.

(3) Anion-water hydrogen bond vibration mode. In NaCl solution, the addition of electronegative chloride ions forms OH···Cl$^-$ hydrogen bonds with water molecules. This hydrogen bond is generally considered to have a vibration associated with the intermolecular hydrogen bond stretching mode [34, 38–40] in the THz frequency range. Based on Raman spectroscopy, the measured OH···Cl$^-$ hydrogen bond vibration mode had a characteristic frequency of ~185 cm^{-1} [40]. The optical Kerr effect (OKE) revealed a characteristic frequency of ~168 cm^{-1} for OH···Cl$^-$ [38]. Although the values are slightly different, the increased positive polarity TKE response can be attributed to this vibration mode. The contribution of the anion-water hydrogen bond vibration mode to the polarizability anisotropy is similar to that of the water-water hydrogen bond stretching mode. This mode has a positive polarity response because the THz electric field is coupled parallel to the direction of the hydrogen bond. The hydrogen bond harmonic oscillator model was used to simulate the TKE response of the OH···Cl$^-$ hydrogen bond vibration mode. The simulation results are presented in Fig. 9.13(d). For the 4 M NaCl solution, the obtained damping coefficient and inherent frequencies were 108 cm^{-1} and 156 cm^{-1}, respectively. This shows

that as the concentration increased, the positive polarity TKE response under the OH···Cl⁻ hydrogen bond vibration mode significantly increased, accompanied by the strengthening of oscillation characteristics.

9.3.2 Influence of Anions

The TKE responses of a series of aqueous halide solutions with different anions were measured and simulated to demonstrate the effectiveness of the proposed model. Fig. 9.14 shows the TKE responses for 4 M solutions of NaCl, sodium bromide (NaBr), and sodium iodide (NaI). For the same concentration, the positive TKE response amplitude evidently increases by order of Cl⁻ < Br⁻ < I⁻, along with the decrease of surface charge density of anion. In particular, the NaI solution exhibits a higher anisotropic response at the same concentration. This is because for anions with low surface charge densities, such as I⁻, the formed OH···I⁻ hydrogen bond has a longer bond length and lower binding energy than the OH···Cl⁻ and OH···Br⁻ hydrogen bonds. In addition, the center frequency of the THz pulse used is 3.9 THz, which is closer to the inherent frequency of OH···I⁻ hydrogen bond vibrations. Therefore, an enhanced dipole moment and greater polarizability anisotropy under the OH···I⁻ hydrogen bond vibration mode can be achieved with resonant THz electric field excitation.

The TKE response of the aqueous halide ion solution is fitted using the proposed model. Figs. 9.15(a), (c), and (e) show the simulated responses

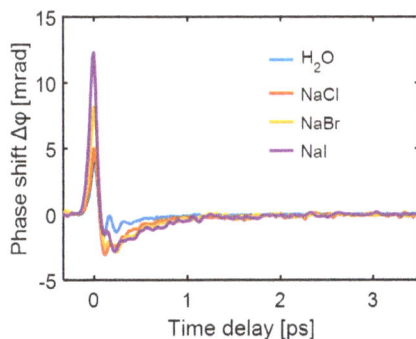

▲ Fig. 9.14.   TKE responses of 4 M NaCl, NaBr, and NaI solutions. Reprinted from Ref. [41] with permission.

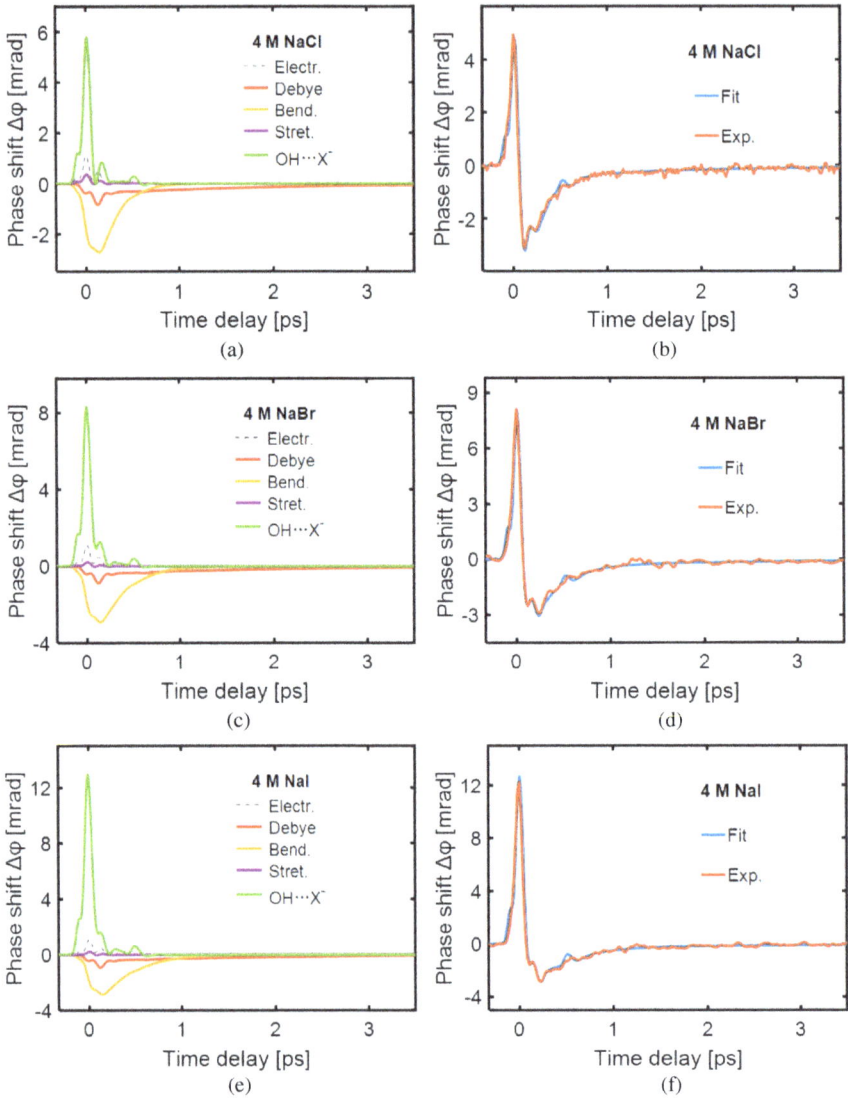

▲ Fig. 9.15. Theoretical simulation of the TKE responses of 4 M (a) NaCl, (b) NaBr, and (c) NaI solutions under the electronic, Debye relaxation, hydrogen bond bending, hydrogen bond stretching, and OH···X⁻ hydrogen bond vibration modes. The red and blue lines in (b), (d), and (f) represent the measured data and the sum of the simulated responses in different modes, respectively. Reprinted from Ref. [36] with permission.

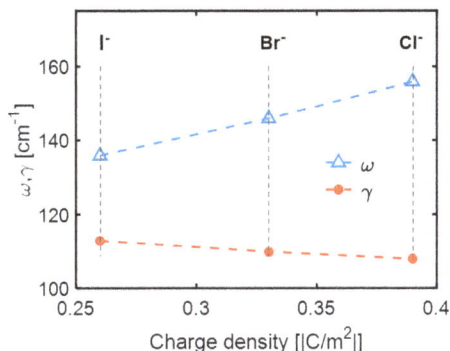

▲ Fig. 9.16.   Inherent frequency ω and damping coefficient γ in NaCl, NaBr, and NaI solutions. Reprinted from Ref. [36] with permission.

of the NaCl, NaBr, and NaI aqueous solutions, respectively, including the electronic, Debye relaxation, hydrogen bond bending and stretching modes, and ion-water hydrogen bond vibration mode. The blue lines in Figs. 9.15(b), (d), and (f) indicate the sum of simulated responses of the different modes, which agree with the experimental data (red lines).

The simulated inherent frequencies and damping coefficients for the OH⋯X$^-$ (X$^-$: Cl$^-$, Br$^-$, I$^-$) hydrogen bond vibration modes, based on the TKE technique, are shown in Figs. 9.16. The obtained inherent frequencies are ~156 cm^{-1}, ~146 cm^{-1}, and ~136 cm^{-1} for the OH⋯Cl$^-$, OH⋯Br$^-$, and OH⋯I$^-$ hydrogen bond vibration modes, respectively. The corresponding damping coefficients were ~108 cm^{-1}, ~110 cm^{-1}, and ~113 cm^{-1}, respectively. The corresponding exponential relaxation time constants were ~98 fs, ~97 fs, and ~94 fs, respectively. The above parameter values are consistent with previous results measured using nonresonant OKE spectroscopy [33].

9.3.3 Effect of Cations

Fig. 9.17(a) shows the TKE responses for different concentrations of MgCl$_2$ aqueous solutions. The blue line is the TKE response of pure water, which is dominated by intermolecular hydrogen bond motion. Specifically, the positive polarity response with significant oscillation information mainly

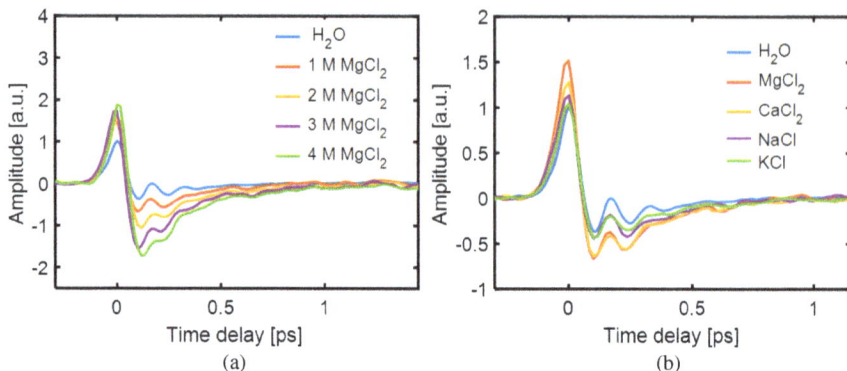

▲ Fig. 9.17.   (a) TKE responses of $MgCl_2$ aqueous solutions with different concentrations. (b) TKE responses of NaCl, KCl, $CaCl_2$, and $MgCl_2$ aqueous solutions with a concentration of 1 M. Reprinted from Ref. [41] with permission.

originates from the intermolecular hydrogen bond stretching vibration mode, and the negative response originates from the coordination of the Debye relaxation process and the intermolecular hydrogen bond bending vibration mode. The induced anisotropic TKE response under intermolecular hydrogen bond bending and stretching modes can be explained using the proposed hydrogen bond harmonic oscillator model. The TKE responses of $MgCl_2$ solutions increase with increasing concentration, which is similar to the results for NaCl solutions. However, at a high concentration of 4 M, the signal vibration characteristics of the $MgCl_2$ solution became noticeable. The interaction between Na^+ and water molecules is equivalent to that between adjacent water molecules [30]. However, the interaction between other cations, such as Mg^{2+}, and water molecules cannot be ignored, complicating the theoretical analysis. Therefore, a low concentration of 1 M is commonly used in the following discussion to exclude the effects of solvent-solute and solute-solute interactions.

Fig. 9.17(b) shows the TKE responses of NaCl, KCl, $CaCl_2$, and $MgCl_2$ aqueous solutions with the same concentrations of 1 M and shows that all the TKE responses are increased compared to those of pure water. The amplitude of the TKE response of water affected by cations is arranged in the order of $K^+ < Na^+ < Ca^{2+} < Mg^{2+}$, which is consistent with the order of surface charge density, that is, K^+ (0.669 C/m^2) $< Na^+$ (1.224 C/m^2) $< Ca^{2+}$ (2.54 C/m^2) $< Mg^{2+}$ (4.91 C/m^2). In general, the oxygen atoms in water molecules close to the cations are fixed under the action of positive charges, thereby enhancing the structure of the water network. We infer that this process promotes the energy transfer between the permanent dipole moment rotation of water molecules and the restricted translational motion of adjacent water molecules with THz pulse excitation, resulting in a significant increase in the bipolar response.

Notably, the negative responses of $CaCl_2$ and $MgCl_2$ aqueous solutions differed from those of NaCl and KCl aqueous solutions, which may be related to the effect of excessive Cl^-. Fig. 9.18(a) shows the TKE responses of 1 M aqueous solutions of $CaCl_2$ and $MgCl_2$ and 2 M aqueous solutions of NaCl and KCl. Even though the molar concentrations of K^+ and Na^+ are twice those of Ca^{2+} and Mg^{2+}, the increase in TKE response amplitude is still arranged in the order of $K^+ < Na^+ < Ca^{2+} < Mg^{2+}$. This implies that the effect of cations on the TKE response is smaller than that of anions. Fig. 9.18(b) shows the TKE responses of 1 M Na_2SO_4 and $MgSO_4$ aqueous solutions. It can be observed that the influence of Mg^{2+} on the TKE signal of water is greater than that of Na^+, even though

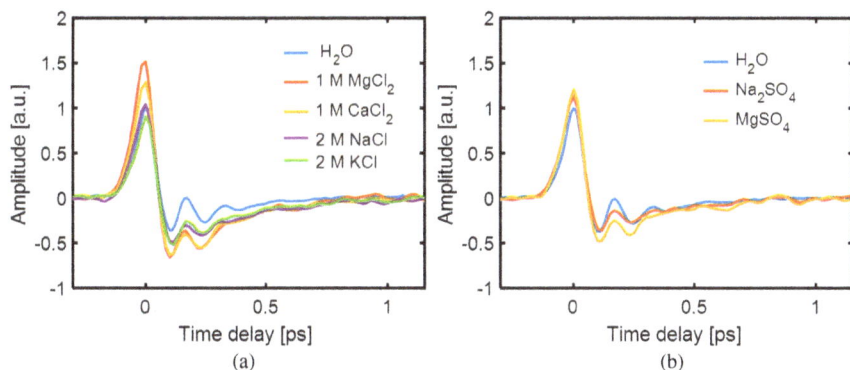

▲ Fig. 9.18. (a) TKE responses of 1 M $CaCl_2$ and $MgCl_2$ aqueous solutions and 2 M NaCl and KCl aqueous solutions. (b) TKE responses of 1 M Na_2SO_4 and $MgSO_4$ aqueous solutions. (b) is reprinted from Ref. [41] with permission.

the molar concentration of Na^+ is twice that of Mg^{2+}, implying that the THz electric field energy is more likely to be coupled to the rotation and restricted translational motion of water molecules under the influence of strongly hydrated ions, such as in the $MgSO_4$ solution.

Evidence suggests that aqueous solutions of Mg^{2+} with a higher surface charge density produce stronger anisotropy with THz electric field excitation, and this ion is recognized as strongly hydrated. In addition, the relaxation timescales of the different cation aqueous solutions are almost the same. This is consistent with the phenomenon observed in dielectric relaxation spectroscopy, which is attributed to the rapid, small angular motion of a water molecule [42]. The fitted average time constants are in the range of 1.28 ± 0.2 ps and, in all cases, are slower than the time constant of ~1.1 ps in pure water. The slower kinetics may be related to the increase in the viscosity of aqueous solutions [32, 33].

9.4 Kerr Effect of Ethanol-Water Mixtures Driven by THz Field

Ethanol is a classic hydrogen-bonded liquid that has attracted extensive attention from researchers. In particular, as the simplest chemically amphiphilic aqueous solutions, alcohol-water mixtures are considered as the reference for exploring molecular interactions in biological processes [43–45]. Here, a high-frequency THz pulse with a center frequency of 3.9 THz is used to excite the ethanol and ethanol-water mixed solutions with different molar proportions and observe the time-resolved evolution of the transient birefringence. The potential microscopic origins that contributed to the polarizability anisotropy were analyzed. For ethanol-water mixtures with different molar proportions, we extracted the concentration-related contribution coefficients of pure ethanol and water on the sub-picosecond timescale and the relative amplitudes of the molecular motions under the observation time window of tens of picoseconds (after 1 ps). In addition, the relative molecular contribution of ethanol in the ethanol-water mixture exhibits a maximum value at a certain molar proportion, which is caused by the destruction of the structure related to the hydrogen bond chain of ethanol. The application of TKE technology in an ethanol-water mixture can provide a basic reference for understanding the interactions of different hydrogen-bonded liquids and facilitate the development of a theoretical model of hydrogen bond dynamics.

9.4.1 Experimental Results and Theoretical Analysis of TKE Signal of Ethanol

The liquid samples in this experiment were ethanol and ethanol-water mixtures. The thickness of the liquid films was 90 ± 4 µm by controlling the flow-regulating valve in a free-flowing liquid film system. Fig. 9.19(a) shows the TKE signals of pure ethanol with different THz pump intensities. The measured unipolar TKE signal of ethanol exhibits two characteristics: (i) a sharp rise at the sub-picosecond timescale, and (ii) a slow signal decay process that extends over tens of picoseconds. Usually, the characteristic (i) of the sharp rise comes from the electronic response of the liquid itself, which is introduced by the electron cloud distortion under THz pump pulse excitation. Therefore, the temporal response of the electronic contribution at the sub-picosecond timescale should follow the THz intensity curve under

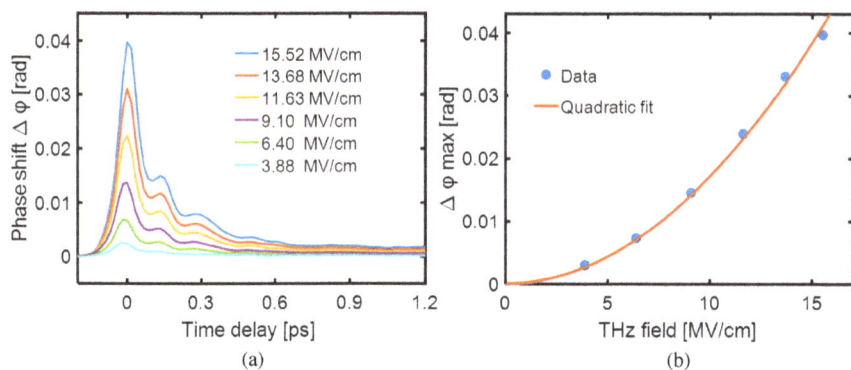

▲ Fig. 9.19.   (a) TKE responses of ethanol with different pump intensities. (b) Intensity dependence curve of the TKE response for ethanol. Reprinted from Ref. [46] with permission.

ideal conditions. The apparent decay of characteristic (ii) was dominated by the molecular motion of ethanol. In addition, the recorded peak values of the TKE signals for different pump intensities are proportional to the square of the pump electric field strength, as shown in Fig. 9.19(b), proving that the measured signals of pure ethanol are dominated by the TKE.

The temporal response of the electronic contribution at the sub-picosecond timescale should follow the THz intensity curve under ideal conditions. This can be expressed as $\Delta n_e = \lambda B_e |E(z,t)|^2$. The relationship between the phase shift of the probe beam at 800 nm and the refractive index anisotropy was established using the following formula: $\Delta\varphi_e(t) = \frac{2\pi}{\lambda} \int_0^l \Delta n_e(z,t)dz$, where λ is the wavelength of the probe beam, and l is the liquid thickness. Considering the inevitable error factors in the sampling process, such as the sampling error caused by the 50 fs probe pulse [47], the phase mismatch between the THz pulse and the 800 nm probe pulse [48, 49], surface reflection [24], and Gouy phase shift during the propagation of the focused THz wave in the ethanol film [24, 27], the electron response in ethanol can be extended to the following formula:

$$\Delta\varphi_e(t) = 2\pi B_e \int_0^l \left[t_{12} E\left(t + \beta x - \arctan(n_{THz}x/Z_R)/\omega_{THz})\right) \cdot \exp(-\alpha x)\right]^2$$

$$\times dx * \exp\left(-4\ln 2\left(\frac{t}{\tau}\right)^2\right),$$

$$(9.5)$$

where $B_e \approx 0.0093 \times 10^{-14}$ m/V^2 represents the electron response coefficient in ethanol [28]. $t_{12} = 2/(n_{THz} + 1)$ denotes the Fresnel transmission coefficient, and $n_{THz} \approx 1.47$ is the refractive index of the THz wave in ethanol. $\beta \approx 0.53$ ps/mm is the phase mismatch factor between the THz pulse and the 800 nm probe pulse. $\exp(-\alpha x)$ represents the attenuation of the THz wave propagated in the sample (for ethanol, $\alpha \approx 140$/cm [50]). $\arctan(n_{THz}x/Z_R)$ denotes the phase shift term and Z_R is the Rayleigh length (~ 0.26 mm in this work). For broadband THz pulses, this phase factor is added to the time term of the THz electric field, which can be obtained by combining the derivative of the THz electric field with the frequency-domain correction

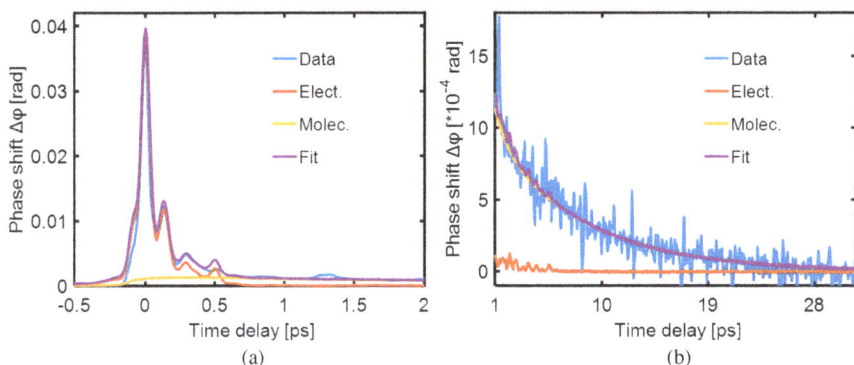

▲ Fig. 9.20. (a) Measured TKE response of ethanol and theoretical simulation results with electronic and molecular responses, respectively. The red line is the sum of the above two contributions, which matches well with the experimental data (blue line). (b) TKE response of ethanol under a long observation time window, which is accompanied by a recovery process extending over tens of picoseconds. Reprinted from Ref. [46] with permission.

and trigonometric operations in the numerical calculation. $\exp(-4\ \ln 2(t/\tau)^2)$ represents the sampling pulse, and τ is the pulse width of the probe pulse (~50 fs in this study). Here, we ignore the frequency dependence of these factors to reduce the large amount of calculation and use the optical parameters at the center frequency of 3.9 THz as the average value for approximation. Considering the above factors, the simulated electronic responses are shown as red lines in Fig. 9.20(a) and are consistent with the measured TKE response under the sub-picosecond observation time window.

Moreover, the frequency spectrum of the applied broadband THz pulse covers various molecular motion modes of ethanol. The apparent decay characteristic (ii) of the measured TKE signal in pure ethanol was dominated by the molecular motions of ethanol. The coupling between the permanent molecule dipole moment of ethanol and the THz electric field can be assigned to two fast Debye relaxation modes, which are related to the formation and breakage of hydrogen bonds with a relaxation time $\tau_1 = 1$–2 ps (D1 process), and the fluctuations of single terminal monomers of the hydrogen bond chain structure with a relaxation time $\tau_2 = 7$–12 ps (D2 process) [43]. A double exponential decay model [51] was used to describe

the molecular contributions to the TKE response of ethanol, as shown by the yellow lines in Fig. 9.20(b). Fitted time constants of $\tau_1 = 1.1$ ps and $\tau_2 = 8$ ps were adopted. Moreover, the sum (purple curve) of the simulated electronic and molecular responses agree with the measured TKE signal (blue). This indicates that the electronic contribution is dominant in the ultrafast evolution process at the sub-picosecond timescale, and the molecular contribution plays a significant role over tens of picoseconds.

9.4.2 Simulation and Analysis of TKE Signals of Ethanol-Water Mixtures

Based on the above TKE measurement results of ethanol, we investigated the TKE responses of the ethanol-water mixtures with different molar proportions under THz pulse excitation, as shown in Fig. 9.21. The mixing ratio of the two liquids is expressed by the molar concentration C of the ethanol molecules, namely $C = 100$ mol% pure ethanol and $C = 0$ mol% pure water. The measured responses have (i) a sharp rise at the sub-picosecond timescale and (ii) a slow signal decay process that extends over tens of picoseconds. To observe the decay characteristics, we normalized the measured TKE response of the ethanol-water mixtures with molar concentrations of 100, 70, and 40 mol% according to the amplitude at $t = 5$

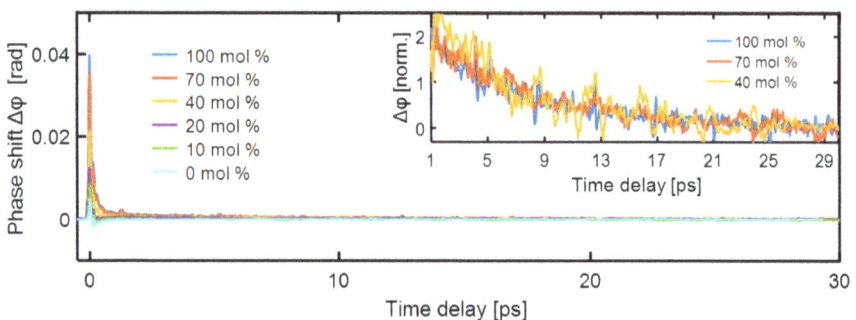

▲ Fig. 9.21.   TKE responses of ethanol-water mixtures with different molar ratios of ethanol/water molecules. The inset shows the normalized TKE response at $t = 5$ ps for the mixtures with molar concentrations $C = 100$ mol%, 70 mol%, and 40 mol%. Reprinted from Ref. [46] with permission.

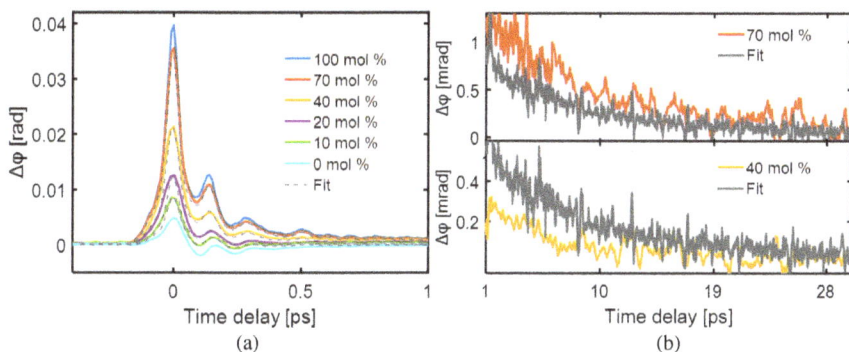

▲ Fig. 9.22. (a) The fitted (gray lines) and measured TKE responses of ethanol-water mixtures with different molar ratios under the sub-picosecond observation time window (before 1 ps). (b) The measured TKE responses of ethanol-water mixtures with concentrations of $C = 70$ mol%, 40 mol% and the corresponding calculated results (gray lines) under the observation time window over tens of picoseconds (after 1 ps). Reprinted from Ref. [46] with permission.

ps (as shown in the inset of Fig. 9.21). The TKE responses for the ethanol-water mixtures with different molar concentrations almost overlap after ~5 ps, indicating that the molecular responses of the mixtures are still dominated by the two fast Debye relaxation modes of ethanol.

To observe the characteristics of the electronic response on the sub-picosecond timescale, the TKE response within 1 ps is shown as a solid line in Fig. 9.22(a). To analyze the relative effects of ethanol and water molecules in the mixture, the TKE responses of ethanol-water mixtures with different molar concentrations were fitted using Eq. (9.6), and the corresponding fitted results are shown by gray lines.

$$\Delta\varphi(t) = a_i \Delta\varphi_{water}(t) + b_i \Delta\varphi_{ethanol}(t), \tag{9.6}$$

where $\Delta\varphi_{water}(t)$ represents the measured TKE response of pure water. The bipolar response of water has been introduced and assigned to the electric, Debye relaxation, intermolecular hydrogen bond stretching vibration, and bending vibration contributions. In particular, transient birefringence is mainly attributed to two modes of intermolecular hydrogen bond

motion. $\Delta\varphi_{ethanol}(t)$ denotes the measured TKE response of ethanol (C = 100 mol%), which can be approximately 10 times greater than that of water under the same conditions. a_i and b_i are the coefficients related to the molar concentration of ethanol. Fitted curves of the TKE responses in ethanol-water mixtures were obtained by varying the corresponding a_i and b_i. Under an ultrafast time window, the contributions of water and ethanol to the TKE response depend on the number of two types of excited molecules, which are mainly related to the molar concentration of ethanol and the THz absorption coefficient of the ethanol-water mixture. Therefore, the fitted TKE responses of the ethanol-water mixtures with different molar concentrations are presented in Fig. 9.22(a), which agree with the measured TKE response. This indicates that the temporal evolution curves of the TKE responses of the mixtures can always be represented by a linear superposition of the TKE responses of pure water and ethanol.

However, if this linear superposition simulation is extended to the timescale of tens of picoseconds (after 1 ps), the measured results with different molar concentrations deviate from the expectations. For example, as shown in Fig. 9.22(b), the measured TKE response of the ethanol-water mixture at C = 70 mol% (red line) is greater than the calculated value (gray line), whereas the measured TKE response at C = 40 mol% (yellow line) is smaller than the calculated value (gray line). A similar deviation from the linear superposition simulation has also been observed in the Raman spectrum study of ethanol-water mixtures [52], implying that there are complex interactions related to the molecular structures in the mixture.

We discuss the TKE responses under the observation time windows of sub-picoseconds and tens of picoseconds separately to further investigate the potential molecular motions. On the sub-picosecond timescale, the fitted values of a_i and b_i are depicted in Fig. 9.23(a). As the number of water molecules increased, the contribution of water to the TKE response of the mixture gradually increased, indicating that the restricted translational motion of adjacent water molecules contributed incrementally to the anisotropic response with the THz pulse excitation. A small number of ethanol molecules in the mixture hardly affected the hydrogen bond

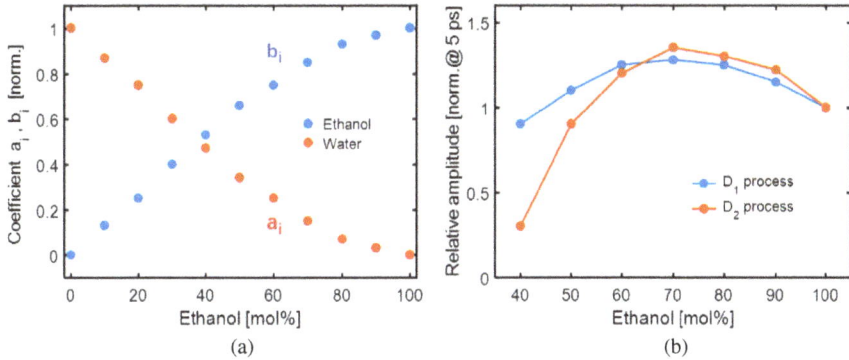

▲ Fig. 9.23. (a) Coefficients a_i and b_i for different molar concentrations under the sub-picosecond observation time window. (b) The relative amplitudes of two Debye relaxation processes at different molar concentrations compared to pure ethanol. Reprinted from Ref. [46] with permission.

network of liquid water. Even with the addition of a large number of ethanol molecules, the TKE response of the mixture still contains anisotropy information about the water-water intermolecular hydrogen bond motion modes, implying that the water intermolecular hydrogen bond structure can exist in the ethanol-rich region. These measurements under the sub-picosecond observation time window are consistent with the results of the low-frequency Raman studies for the ethanol-water mixture, which show that the intermolecular vibrational modes are independent of each other in the mixture of these two liquids [52].

Under the observation time window of tens of picoseconds, we used a double exponential decay model to simulate the relaxation processes in the ethanol-water mixtures with different molar concentrations. This fitting method has been proven appropriate in recent molecular dynamics studies.

$$\Delta\varphi_m\left(t\right) \propto M\left(1 - A_{c=i}/\tau_1 \exp(-t/\tau_1) - B_{c=i}/\tau_2 \exp(-t/\tau_2)\right) \quad (9.7)$$

On the premise that the fitted curve and the measured data are in good agreement, the same values of time constants ($\tau_1 = 1.1$ ps and $\tau_2 = 8$ ps) were used for the mixtures with different molar concentrations, and the amplitudes of $A_{c=i}$, $B_{c=i}$ related to the two molecular relaxation modes were

recorded ($C = i$ represents the ethanol molar concentration of the ethanol-water mixture). To more directly quantify and analyze the deviations in the mixtures of different molar concentrations, the recorded amplitudes of $A'_{C=i}$, $B'_{C=i}$ were divided by a concentration-related linear fitting value to obtain the values of $A'_{C=i} = A_{C=i}/A_{C=1}b_i$, $B'_{C=i} = B_{C=i}/B_{C=1}b_i$. b_i represents the relative amplitudes of the two Debye relaxation processes at different molar concentrations compared with those of pure ethanol. As shown in Fig. 9.23(b), the blue line represents the relative amplitude $A'_{C=i}$ for the D_1 relaxation process with different molar concentrations. When adding a small number of water molecules to ethanol, we observed a relatively enhanced D_1 relaxation process, which was due to the increased molecular motions associated with hydrogen bond breakage and the formation of ethanol. However, as the number of water molecules increased, the relative amplitude of the D_1 relaxation process decreased, and an inflection point appeared at approximately 70 mol%.

The relative amplitude $B'_{C=i}$ of the D_2 relaxation process depicted by the red line in Fig. 9.23(b) is associated with the fluctuations of terminal ethanol monomers of the hydrogen bond chain. When a small amount of water was introduced into ethanol, the relative amplitude increased as the molar concentration of ethanol decreased in the range of $C = 100$–70 mol%. However, as the number of water molecules increased, the relative amplitude gradually decreased to 70–30 mol% and quickly dropped to the noise level when C was less than ~30 mol%. The addition of a small number of water molecules (in the range of $C = 100$–70 mol%) causes the long hydrogen bond chain of ethanol to break into several short chains, thereby increasing the number of terminal ethanol monomers of the chains and the probability of the formation and breakage of ethanol intermolecular hydrogen bonds [45]. When water molecules were added (in the range of $C = 70$–30 mol%), the number of free ethanol monomers isolated by water molecules gradually increased, resulting in a decrease in the proportion of terminal ethanol monomers in the chain. When the molar concentration of ethanol was reduced to below 30 mol%, the amplitude of the D_2 relaxation process decreased to the noise level. This is primarily because ethanol molecular clusters are gradually dissolved by water molecules, which makes it challenging to maintain the chain-like structure of ethanol.

9.5 Conclusion

THz pulses with peak intensities above the order of MV/cm can be successfully obtained with the development of THz technology. Such a high electric field of THz wave opens up a new field of nonlinear THz spectroscopy, enabling resonant excitation of the low-frequency motion of matter. An enhanced molecular response can be obtained using a THz electric field as the field resonates with the rotational transitions of a single molecule or cooperative low-frequency molecular motions. Therefore, the time-resolved TKE response is expected to be a powerful phenomenon for exploring low-frequency dynamics of materials.

This chapter uses TKE techniques to observe the low-frequency molecular collective/cooperative motion associated with hydrogen bonding in liquid water, combined with kinetic analysis, to explain the microscopic molecular motion mechanism revealed by the TKE response. On this basis, the exploration of low-frequency dynamics has been extended to more complex situations by adding a variety of ionic and molecular solutes. The interactions between the ionic/molecular solutes and the hydrogen bond network of water molecules were further analyzed. This research provides a basis for further exploration of energy coupling and transfer in the hydrogen bond network of liquids and the interaction of molecules, ions, and even biological macromolecules in aqueous solutions. We hope this study provides a useful reference for further insights into the effects of hydrogen bond networks on biochemical reactions in aqueous environments in the future.

References

1. Omta A. W., Kropman M. F., Woutersen S. & Bakker H. J. (2003). Negligible effect of ions on the hydrogen-bond structure in liquid water, Science, 301(5631), pp. 347–349.

2. Fecko C. J., Eaves J. D., Loparo J. J., Tokmakoff A. & Geissler P. L. (2003). Ultrafast hydrogen-bond dynamics in the infrared spectroscopy of water, Science, 301(5640), pp. 1698–1702.

3. Smith J. D., Cappa C. D., Wilson K. R., Messer B. M., Cohen R. C. & Saykally R. J. (2004). Energetics of hydrogen bond network rearrangements in liquid water, Science, 306(5697), pp. 851–853.

4. Stiopkin I. V., Weeraman C., Pieniazek P. A., Shalhout F. Y., Skinner J. L. & Benderskii A. V. (2011). Hydrogen bonding at the water surface revealed by isotopic dilution spectroscopy, Nature, 474(7350), pp. 192–195.

5. Richardson J. O., Pérez C., Lobsiger S., Reid A. A., Temelso B., Shields G. C., Kisiel Z., Wales D. J., Pate B. H. & Althorpe S. C. (2016). Concerted hydrogen-bond breaking by quantum tunneling in the water hexamer prism, Science, 351(6279), pp. 1310–1313.

6. Stokely K., Mazza M. G., Stanley H. E. & Franzese G. (2010). Effect of hydrogen bond cooperativity on the behavior of water, Proceedings of the National Academy of Sciences, 107(4), pp. 1301–1306.

7. Sharma M., Resta R. & Car R. (2005). Intermolecular dynamical charge fluctuations in water: A signature of the H-bond network, Physical Review Letters, 95(18), p. 187401.

8. Heyden M. & Tobias D. J. (2013). Spatial dependence of protein-water collective hydrogen-bond dynamics, Physical Review Letters, 111(21), p. 218101.

9. Bakker H. J. & Skinner J. L. (2010). Vibrational spectroscopy as a probe of structure and dynamics in liquid water, Chemical Reviews, 110(3), pp. 1498–1517.

10. Kühne T. D., Krack M. & Parrinello M. (2009). Static and dynamical properties of liquid water from first principles by a novel Car–Parrinello-like approach, Journal of Chemical Theory and Computation, 5(2), pp. 235–241.

11. Kampfrath T., Wolf M. & Sajadi M. (2017). Anharmonic coupling between intermolecular motions of water revealed by terahertz Kerr effect, arXiv preprint arXiv:1707.07622.

12. Elgabarty H., Kampfrath T., Bonthuis D. J., Balos V., Kaliannan N. K., Loche P., Netz R. R., Wolf M., Kühne T. D. & Sajadi M. (2020). Energy transfer within the hydrogen bonding network of water following resonant terahertz excitation, Science Advances, 6(17), p. eaay7074.

13. Perakis F., De Marco L., Shalit A., Tang F., Kann Z. R., Kühne T. D., Torre R., Bonn M. & Nagata Y. (2016). Vibrational spectroscopy and dynamics of water, Chemical Reviews, 116(13), pp. 7590–7607.

14. Vij J. K., Simpson D. R. J. & Panarina O. E. (2004). Far infrared spectroscopy of water at different temperatures: GHz to THz dielectric spectroscopy of water, Journal of Molecular Liquids, 112(3), pp. 125–135.

15. Torii H. (2011). Intermolecular electron density modulations in water and their effects on the far-infrared spectral profiles at 6 THz, The Journal of Physical Chemistry B, 115(20), pp. 6636–6643.

16. Mizoguchi K., Hori Y. & Tominaga Y. (1992). Study on dynamical structure in water and heavy water by low-frequency Raman spectroscopy, The Journal of Chemical Physics, 97(3), pp. 1961–1968.

17. Fukasawa T., Sato T., Watanabe J., Hama Y., Kunz W. & Buchner R. (2005). Relation between dielectric and low-frequency Raman spectra of hydrogen-bond liquids, Physical Review Letters, 95(19), p. 197802.

18. Rønne C. & Keiding S. R. (2002). Low frequency spectroscopy of liquid water using THz-time domain spectroscopy, Journal of Molecular Liquids, 101(1–3), pp. 199–218.

19. Penkov N., Shvirst N., Yashin V., Fesenko E, Jr. & Fesenko E. (2015). Terahertz spectroscopy applied for investigation of water structure, The Journal of Physical Chemistry B, 119(39), pp. 12664–12670.

20. Savolainen J., Ahmed S. & Hamm P. (2013). Two-dimensional Raman-terahertz spectroscopy of water, Proceedings of the National Academy of Sciences, 110(51), pp. 20402–20407.

21. Soper A. K. (2000). The radial distribution functions of water and ice from 220 to 673 K and at pressures up to 400 MPa, Chemical Physics, 258(2–3), pp. 121–137.

22. Laage D. (2009). Reinterpretation of the liquid water quasi-elastic neutron scattering spectra based on a nondiffusive jump reorientation mechanism, The Journal of Physical Chemistry B, 113(9), pp. 2684–2687.

23. Teixeira J., Bellissent-Funel M.-C., Chen S. H. & Dianoux A. J. (1985). Experimental determination of the nature of diffusive motions of water molecules at low temperatures, Physical Review A, 31(3), p. 1913.

24. Sajadi M., Wolf M. & Kampfrath T. (2015). Terahertz-field-induced optical birefringence in common window and substrate materials, Optics Express, 23(22), pp. 28985–28992.

25. Leitenstorfer A., Hunsche S., Shah J., Nuss M. C. & Knox W. H. (1999). Detectors and sources for ultrabroadband electro-optic sampling: Experiment and theory, Applied Physics Letters, 74(11), pp. 1516–1518.

26. Wu Q. & Zhang X.-C. (1997). 7 terahertz broadband GaP electro-optic sensor, Applied Physics Letters, 70(14), pp. 1784–1786.

27. Bakker H. J., Cho G. C., Kurz H., Wu Q. & Zhang X.-C. (1998). Distortion of terahertz pulses in electro-optic sampling, JOSA B, 15(6), pp. 1795–1801.

28. Zalden P., Song L., Wu X., Huang H., Ahr F., Mücke O. D., Reichert J., Thorwart M., Mishra P. Kr., Welsch R., Santra R., Kärtner F. X. & Bressler C. (2018). Molecular polarizability anisotropy of liquid water revealed by terahertz-induced transient orientation, Nature Communications, 9(1), pp. 1–7.

29. Novelli F., Pestana L. R., Bennett K. C., Sebastiani F., Adams E. M., Stavrias N., Ockelmann T., Colchero A., Hoberg C., Schwaab G., Head-Gordon. T. & Havenith M. (2020). Strong anisotropy in liquid water upon librational excitation using terahertz laser fields, The Journal of Physical Chemistry B, 124(24), pp. 4989–5001.

30. Collins K. D. (1995). Sticky ions in biological systems, Proceedings of the National Academy of Sciences, 92(12), pp. 5553–5557.

31. Maroulis G. (1998). Hyperpolarizability of H_2O revisited: accurate estimate of the basis set limit and the size of electron correlation effects, Chemical Physics Letters, 289(3–4), pp. 403–411.

32. Tielrooij K. J., Van Der Post S. T., Hunger J., Bonn M. & Bakker H. J. (2011). Anisotropic water reorientation around ions, The Journal of Physical Chemistry B, 115(43), pp. 12638–12647.

33. Heisler I. A., Mazur K. & Meech S. R. (2011). Low-frequency modes of aqueous alkali halide solutions: An ultrafast optical Kerr effect study, The Journal of Physical Chemistry B, 115(8), pp. 1863–1873.

34. Zhao H., Tan Y., Zhang L., Zhang R., Shalaby M., Zhang C., Zhao Y. & Zhang X.-C. (2020). Ultrafast hydrogen bond dynamics of liquid water revealed by terahertz-induced transient birefringence, Light: Science & Applications, 9(1), pp. 1–10.

35. Demtröder W. (2008). *Laser Spectroscopy: Vol. 2 Experimental Techniques*. José María Aguirre Oraa, pp. 77–104.

36. Zhao H., Tan Y., Zhang R., Zhao Y., Zhang C. & Zhang L. (2021). Anion-water hydrogen bond vibration revealed by the terahertz Kerr effect, Optics Letters, 46(2), pp. 230–233.
37. Wang Y. & Tominaga Y. (1994). Dynamical structure of water in aqueous electrolyte solutions by low-frequency Raman scattering, The Journal of Chemical Physics, 101(5), pp. 3453–3458.
38. Heisler I. A. & Meech S. R. (2010). Low-frequency modes of aqueous alkali halide solutions: glimpsing the hydrogen bonding vibration, Science, 327(5967), pp. 857–860.
39. Wu D. Y., Duan S., Liu X. M., Xu Y. C., Jiang Y. X., Ren B., Xu X., Lin S. H. & Tian Z. Q. (2008). Theoretical study of binding interactions and vibrational Raman spectra of water in hydrogen-bonded anionic complexes:(H_2O) n- (n = 2 and 3), $H_2O\cdots X^-$ (X = F, Cl, Br, and I), and $H_2O\cdots M^-$ (M = Cu, Ag, and Au), The Journal of Physical Chemistry A, 112(6), pp. 1313–321.
40. Craig J. D. C. (2002). Raman spectroscopic and calculated vibrational wavenumbers of anion hydrates, Journal of Raman Spectroscopy, 33(3), pp. 191–196.
41. Zhao H., Tan Y., Wu T., Zhang R., Zhao Y., Zhang C. & Zhang L. (2021). Strong anisotropy in aqueous salt solutions revealed by terahertz-induced Kerr effect, Optics Communications, 497, p. 127192.
42. Balos V., Kim H., Bonn M. & Hunger J. (2016). Dissecting Hofmeister effects: direct anion–amide interactions are weaker than cation–amide binding, Angewandte Chemie International Edition, 55(28), pp. 8125–8128.
43. Skaf M. S., Fonseca T. & Ladanyi B. M. (1993). Wave vector dependent dielectric relaxation in hydrogen-bonding liquids: A molecular dynamics study of methanol, The Journal of Chemical Physics, 98(11), pp. 8929–8945.
44. Allen M. P. & Tildesley D. J. (1987). *Molecular Simulation of Liquids.* Oxford University Press.
45. Yomogida Y., Sato Y., Nozaki R., Mishina T. & Nakahara J. I. (2010). Dielectric study of normal alcohols with THz time-domain spectroscopy, Journal of Molecular Liquids, 154(1), pp. 31–35.
46. Zhao H., Tan Y., Zhang R., Zhao Y., Zhang C., Zhang X.-C. & Zhang L. (2021). Molecular dynamic investigation of ethanol-water mixture

by terahertz-induced Kerr effect, Optics Express, 29(22), pp. 36379–36388.

47. Barthel J., Bachhuber K., Buchner R. & Hetzenauer H. (1990). Dielectric spectra of some common solvents in the microwave region: Water and lower alcohols, Chemical Physics Letters, 165(4), pp. 369–373.

48. Cardona J., Sweatman M. B. & Lue L. (2018). Molecular dynamics investigation of the influence of the hydrogen bond networks in ethanol/water mixtures on dielectric spectra, The Journal of Physical Chemistry B, 122(4), pp. 1505–1515.

49. Ahmed S., Savolainen J. & Hamm P. (2014). The effect of the Gouy phase in optical-pump-THz-probe spectroscopy, Optics Express, 22(4), pp. 4256–4266.

50. Zasetsky A. Y., Lileev A. S. & Lyashchenko A. K. (2010). Molecular dynamic simulations of terahertz spectra for water-methanol mixtures, Molecular Physics, 108(5), pp. 649–656.

51. Li R., D'Agostino C., McGregor J., Mantle M. D., Zeitler J. A. & Gladden L. F. (2014). Mesoscopic structuring and dynamics of alcohol/water solutions probed by terahertz time-domain spectroscopy and pulsed field gradient nuclear magnetic resonance, The Journal of Physical Chemistry B, 118(34), pp. 10156–10166.

52. Buchner R. & Barthel J. (1995). Kinetic processes in the liquid phase studied by high-frequency permittivity measurements, Journal of Molecular Liquids, 63(1–2), pp. 55–75.

© 2024 World Scientific Publishing Company
https://doi.org/10.1142/9789811265648_0010

Chapter 10

Liquid-based Coherent Detection of Broadband Terahertz Pulses

Liangliang Zhang, Wen Xiao, Minghao Zhang, Yong Tan, Cunlin Zhang

Key Laboratory of Terahertz Optoelectronics (MoE), Department of Physics, Capital Normal University, China

10.1 Introduction

Both solids and gases have been demonstrated as materials for terahertz (THz) wave coherent detection [1–5]. Gas-based detection methods require high-energy probe laser beams [6, 7], and the detection bandwidth is limited in solid-based methods [8, 9]. Whether liquids can be used for coherent detection of THz waves is of significant interest in the THz community. Therefore, it is valuable to explore the interaction between lasers and liquid media and develop new technologies for THz wave generation and detection. Compared with the gas medium, liquid media have larger molecular density and nonlinear coefficient, which cause liquid plasma to have a larger free-electron concentration and lower ionization threshold [10–12]. Compared with solid media, liquid media have fluidity and comparable molecular density, and their damage threshold is higher and can be self-repaired. In 2017, researchers reported THz generation based on liquid plasma for the first time [13, 14], which provided new opportunities for the development of THz-related devices based on liquid media. Moreover, related experiments on THz radiation generated by near-infrared laser excitation in liquid water have proved the strong nonlinear coupling process between THz waves and near-infrared lasers in liquid water.

In this chapter, we introduced liquid-based coherent detection of broadband THz pulses. The time-resolved waveform of the THz field with the frequency range of 0.1–18 THz was successfully achieved. The required probe laser energy was as low as a few microjoules. The sensitivity was one order of magnitude higher than that of the air-based method under comparable detection conditions. The energy scaling and polarization properties of the THz-induced beam indicated attribution of the underlying mechanism to a four-wave mixing process. In addition, we extended water to other liquids, such as aqueous salt solutions and ethanol. The ethanol- and solution-based coherent detection scheme further improved the detection sensitivity and signal-to-noise (SNR) ratio, and obtained the proportional relationship of nonlinear refractive index of different solutions in the THz band. The liquid-based THz wave coherent detection scheme broadens the variety of THz wave detectors and provides the possibility of revealing the molecular interaction mechanisms in a biological liquid environment.

10.2 Theory of Coherent Detection of THz Waves based on Liquids

10.2.1 Four-Wave Mixing Mechanism

Generally, both gases and liquids are isotropic media, and their physical properties are independent of the measurement direction. The even-order dielectric polarizability tensor is zero; thus, first- and third-order nonlinear processes should be considered. The third-order nonlinear optical effect is dominant during the interaction between laser-induced plasma and THz waves. In early research on THz wave generation from two-color laser-induced plasma, the dominant third-order nonlinear coefficient was $\chi^{(3)}(\omega, \omega_1, \omega_2, \omega_3)$. According to nonlinear optics theory, $\chi^{(3)}$ is not equal to zero only when $\omega = \omega_1 + \omega_2 + \omega_3$. Therefore, based on the four-wave mixing mechanism, the third-order nonlinear coefficient can be denoted as $\chi^{(3)}$ $(\omega_{THz}, 2\omega + \omega_{THz}, -\omega, -\omega)$, where ω_{THz}, 2ω, and ω represent the frequencies of the THz, second harmonic (SH), and fundamental waves, respectively. The THz electric field radiated at the plasma can be described as

$$E_{THz} \propto \chi^{(3)} E_{2\omega} E_{-\omega} E_{-\omega}, \tag{10.1}$$

where $E_{2\omega}(t) = |E_{2\omega}| e^{i(2\omega t + \omega_{THz} t + \varphi)}$ and $E_{-\omega}(t) = |E_{-\omega}| e^{-i\omega t}$ are the electric-field distributions of the SH and fundamental beams, respectively. φ is the phase difference between the SH and fundamental beams. Substituting the electric field of the fundamental and SH beams into Eq. (10.1), the following equation is obtained:

$$E_{THz} \propto |E_{2\omega}| \cdot |E_\omega|^2 \, \mathrm{Re}[\chi^{(3)} e^{i(\omega_{THz} + \varphi)}]. \tag{10.2}$$

The four-wave mixing mechanism is concise in describing the generated THz field in relation to the energy and polarization of the fundamental and SH beams. However, the third-order nonlinear coefficients $\chi^{(3)}$ of air and water are difficult to be accurately obtained. This is especially so for the pulsed fundamental and SH beams, wherein the pulse duration is usually less than 100 fs. The real and imaginary parts of $\chi^{(3)}$ directly affect the relationship between the THz wave generation efficiency and the relative phase between the fundamental and SH beams. Based on the experimental results of Kim et al. in 2009, the THz radiation intensity generated in air plasma exhibited a relationship similar to a sinusoidal function with the phase difference between the fundamental and SH beams [15]. Therefore, it can be speculated that the imaginary part of the third-order nonlinear coefficient of plasma in air may have a larger absolute value. This also demonstrates that the THz wave conversion efficiency of up to 10^{-4} in air plasma does not come from the small real part of $\chi^{(3)}$.

In the process described by the four-wave mixing model, the four laser beams involved were all single frequency. The wideband characteristic of the input pulse provided a possibility for the $2\omega + \omega_{THz}$ frequency component (or $\omega-\omega_{THz}/2$ fundamental component) of the SH beam. In addition,

another limitation of the four-wave mixing model is that it is based on the stability of plasma properties, which means that $\chi^{(3)}$ does not change with the various input optical fields. At a low laser intensity, the relationship between the generated THz wave energy and input fundamental and SH beams is in good agreement with Eq. (10.1). However, when the input laser field is continuously enhanced, the electron density in the plasma becomes saturated, and the plasma exhibits serious clamp and defocusing effects. Consequently, it is difficult to maintain the same ratio coefficient for THz wave energy conversion efficiency. The THz wave output power tends to be constant when the plasma is saturated. In contrast, $\chi^{(3)}$ decreases with an increase in pump energy, which is beyond the scope of the four-wave mixing model.

In general, the detection process of THz waves based on plasma is the inverse of THz wave generation. When the THz pulse and collinear propagation of fundamental beam are focused to form a plasma, two fundamental beam photons and a THz wave photon produce an SH beam photon, which is a THz-induced second harmonic (TISH). The optical field satisfies the following equation:

$$E_{2\omega\_TISH} \propto \chi^{(3)} E_{\omega} E_{\omega} E_{THz},\qquad(10.3)$$

where $E_{2\omega\_TISH}$ is the SH field strength induced by THz waves, and E_{ω} and E_{THz} are the input fundamental and THz fields, respectively. When the fundamental field is an ultrashort near-infrared pulse (pulse duration of 50 fs, central wavelength of 800 nm), the variation within the timescale of the interaction between the THz and laser fields can be ignored and regarded as a direct current field. In this case, the energy of the generated SH wave can be approximated as follows:

$$I_{2\omega\_TISH}(t) \propto [\chi^{(3)} |I_{\omega}|]^2 I_{THz}(t).\qquad(10.4)$$

The phase information of the detected THz pulse is lost, which induces an incoherent detection. The method of THz wave coherent detection using air plasma based on a four-wave mixing mechanism was first proposed by Zhang *et al.* in 2006 [5]. By significantly increasing the energy of the 800 nm fundamental wave to induce air plasma, a local second-harmonic wave (local oscillation, LO) was obtained from the supercontinuum white light of the air plasma. A 400 nm narrow-band filter was used to extract TISH and LO. Therefore, the measured SH signal can be expressed as:

$$
\begin{aligned}
I_{2\omega\_Signal} &\propto \left(E_{2\omega\_TISH} + E_{2\omega\_LO}\right)^2 \\
&= \left(E_{2\omega\_TISH}\right)^2 + \left(E_{2\omega\_LO}\right)^2 + 2E_{2\omega\_TISH}E_{2\omega\_LO}\cos\varphi \qquad (10.5) \\
&\propto \left(\chi^{(3)}I_\omega\right)^2 I_{THz} + \left(E_{2\omega\_LO}\right)^2 + 2\chi^{(3)}I_\omega E_{2\omega\_LO}E_{THz}\cos\varphi
\end{aligned}
$$

where φ is the phase difference between $E_{2\omega\_TISH}$ and $E_{2\omega\_LO}$. Although simplification of the four-wave mixing mechanism in Eq. (10.5) is incomplete, for an un-adjustable E_{LO}, φ is a constant value, and the measurement results agree well with the expectation under the description of Eq. (10.5). For a weak pump beam, $E_{2\omega\_LO}$ can be neglected. In this case, the second and third terms in Eq. (10.5) can also be ignored, therefore the SH signal amplitude becomes proportional to the THz pulse intensity (E_{THz}^2), and the measurement result is unipolar, which leads to incoherent detection. For a very high pump beam energy, $E_{2\omega\_LO}$ is far greater than $E_{2\omega\_TISH}$, and the first term of Eq. (10.5) is negligible. The second term can be excluded from the chopper and lock-in amplifiers. The third term is dominant and the measured signal is proportional to the THz field E_{THz}, which introduces coherent detection.

10.2.2 THz-Induced Second Harmonic Wave in Liquid Water Plasma

An experimental system measuring the SH radiation in liquid water plasma excited by THz waves is shown in Fig. 10.1(a). A vertically polarized broadband THz wave was generated by an organic 4-N,N-dimethylamino-4′-N′-methyl-stilbazoliumtosylate (DAST) crystal under the excitation of a 1550 nm femtosecond laser pulse (repetition rate of 1 kHz, pulse duration of 50 fs). A set of THz low-pass filters (LPF) was placed in the optical path to filter out the residual infrared pulses. A pair of wire-grid

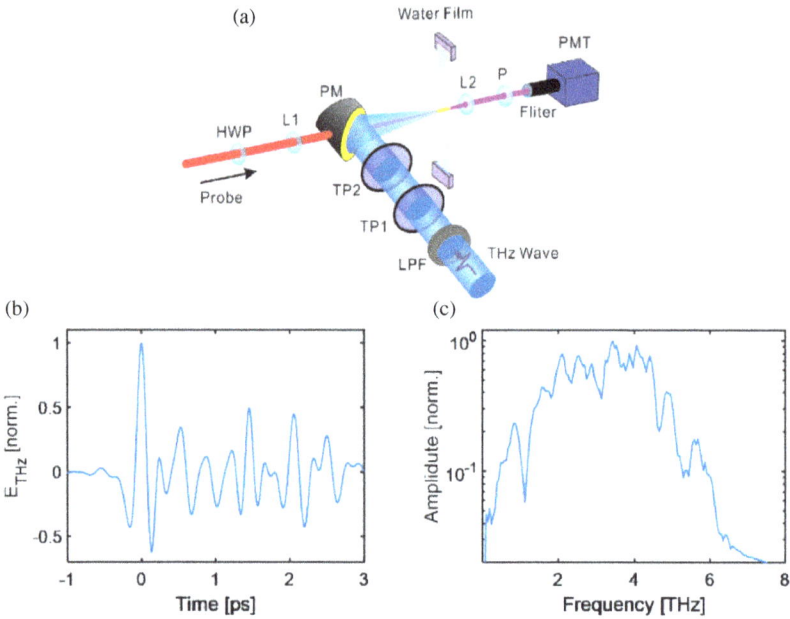

▲ Fig. 10.1. (a) Schematic of TISH measurement system in liquid water. HWP: half-wave plate, L: convex lens, P: 400 nm polarizer, TP: THz polarizer, LPF: THz low-pass filter, Filter: 400 nm band-pass filter. (b) Time-domain spectrum of THz pulses with a cutoff frequency of 6 THz and (c) corresponding spectrum.

THz polarizers was used to control the THz field strength. To obtain the time-domain waveform and its corresponding frequency spectrum shown in Figs. 10.1(b) and (c), we limited the bandwidth of the THz pulse below 6 THz by using a low-pass filter and controlled the THz field strength at 0.3 MV/cm to ensure the detection validity based on the 100 μm thick GaP crystal.

The THz wave was collimated and refocused onto water plasma using a pair of off-axis parabolic mirrors (PM). A probe beam with the same polarization along the vertical direction was co-focused on a 90 μm thick free-flowing water film, and a water plasma was formed by the probe beam above 2 μJ. Then, the 400 nm SH beam, emitted from the water plasma, was measured by a photomultiplier tube (PMT) through a 400 nm narrow-band filter.

The SH LO of the supercontinuum radiation in the air and water plasma was measured when the THz pulse was blocked. To accommodate the focal radius of the THz wave, the water film was placed slightly in front of the 800 nm focal spot to form a beam spot with a diameter of 200 µm. Fig. 10.2(a) and (b) show the SH intensities in air and water plasma, respectively. In this experiment, the gravity-driven water film broke when the single-pulse pump laser energy exceeded ~200 µJ during measurement. In this case, when the energy exceeded the damage threshold of the water film, the water film could hardly be refreshed within the time interval of two adjacent laser pulses (1 ms). Fig. 10.2(a) and (b) show the measurement only when the single-pulse energy was less than 200 µJ. In addition, the inset in Fig. 10.2(a) shows the measurement results excited by higher-energy pulses in the air plasma. The SH component generated in the water plasma was approximately 100 times higher than that in air, which resulted from the large nonlinear coefficient of liquid compared with air.

Based on the system depicted in Fig. 10.1(a), the SH wave in the plasma was greatly enhanced when a THz pulse was employed. When the energy of the probe beam was weak, the TISH radiated from the plasma was much larger than the LO (which can be ignored). By scanning the time delay t_{THz}, a time-domain waveform of the TISH could be obtained. Fig. 10.3 shows

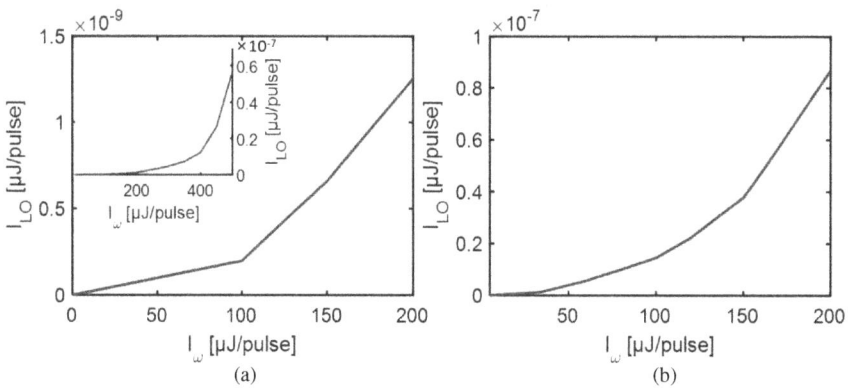

Fig. 10.2. SH components in the supercontinuum radiation of (a) air and (b) water plasmas.

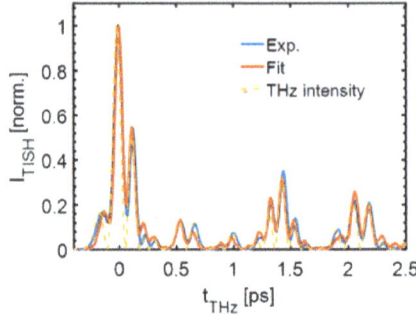

▲ Fig. 10.3.   Time-domain waveform of TISH intensity in water plasma. The orange line is the fitting result based on the four-wave mixing mechanism, and the dashed line represents the intensity of THz field.

the TISH radiation in water plasma. The SH radiation intensity (blue solid line) changed with the THz pulse intensity (yellow dashed line).

Notably, although the generation and detection of THz waves in plasma channels can be analyzed in more detail in the semi-classical plasma photocurrent model, weak plasma can usually be described non-quantitatively by third-order optical rectification. This approximate nonlinear model is important for developing the water and air plasmas into convenient and flexible THz wave generators and detectors. Based on the four-wave mixing mechanism, the THz wave pulse was approximately simplified to a monochromatic wave, and we obtained the following:

$$
\begin{aligned}
E_{2\omega\_TISH}(t) &\propto \mathrm{Re}\Big[\chi^{(3)} A_\omega^2 e^{2i\omega t} A_{THz} e^{i\omega_{THz}t}\Big] \\
&\propto \chi^{(3)} A_\omega^2 A_{THz}\big[\cos 2\omega t \cos \omega_{THz} t + \sin 2\omega t \sin \omega_{THz} t\big] \qquad (10.6) \\
&\propto \chi^{(3)}\left[E_{THz}(t)E_\omega^2(t) + E_{THz}\left(t + \frac{\pi}{2\omega_{THz}}\right)E_\omega^2(t)\right]
\end{aligned}
$$

For a probe pulse $E_\omega(t) = A_\omega \sqrt{\exp(-4\ln 2(t/\tau)^2)} \cos(\omega t)$ with a finite pulse duration (full width at half maximum $= \tau$), in the time-resolved sampling process, the TISH energy generated at each sampling time point t_{THz} represents the result of the interaction between the THz and fundamental fields for the entire duration of the fundamental probe pulse:

$$I_{2\omega\_TISH}(t_{THz}) \propto \int_{-\infty}^{+\infty} \Big[\chi^{(3)} E_\omega^2(t) \big(E_{THz}(\omega_{THz}(t_{THz}+t))$$

$$+ E_{THz}\Big(\omega_{THz}(t_{THz}+t) + \frac{\pi}{2} \Big) \Big) \Big]^2 dt \qquad (10.7)$$

$$\propto \chi^{(3)} A_\omega^4 A_{THz}^2(t_{THz})$$

where A_{THz} represents the amplitude of the THz pulse. Here, the four-wave mixing model was used to analyze the experimental results, and the fitting result (red line) is shown in Fig. 10.3. It can be seen that the measured TISH waveform in water plasma follows the THz wave intensity E_{THz}^2. However, the phase information was completely lost and only the amplitude envelope of the THz pulse could be obtained.

According to Eq. (10.7), the TISH intensity in the liquid-water plasma is proportional to the square of the THz field. The THz field strength was controlled by a pair of THz wire-grid polarizers. By adjusting the THz and probe beam pulses overlap ($t_{THz} = 0$) in time and space, the dependencies of the TISH intensity and THz field strength are shown in Fig. 10.4. The red line represents the quadratic fitting result, which agreed well with the experimental data.

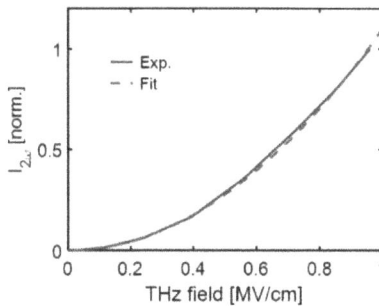

▲ Fig. 10.4. Dependence of TISH intensity as a function of THz field, where a quadratic fit is presented by the red dashed line.

10.3 Coherent Detection of Broadband THz Pulses Based on Liquid Water

Herein, we demonstrated a coherent detection scheme for broadband THz pulses in liquid water [16]. We obtained the temporal waveform of a THz field with a frequency range of 0.1–18 THz by combining a TISH beam generated in water plasma with a control SH (CSH), where we proved the attribution of the underlying mechanism to four-wave mixing. Compared to air-based detection, our proposed water-based scheme was well implemented with a much lower probe laser energy of even a few microjoules, and the detection sensitivity was one order of magnitude higher under comparable conditions. This was the first time a THz coherent detection scheme based on liquid had been realized, which made THz detection more achievable owing to the low probe laser energy requirement. Hence, our scheme is favorable for situations in the absence of sufficiently high probe laser energy, particularly when the probe laser beam needs to propagate a long distance to mix with the THz field and the beam energy is significantly depleted in the optical path.

A schematic of our experimental setup is depicted in Fig. 10.5(a). A vertically polarized broadband intense THz pulse was generated by an organic DAST crystal under the excitation of 1550 nm femtosecond laser pulses. The THz pulse and an 800 nm probe beam with the same polarization along the vertical direction were co-focused onto a 90 µm free-flowing water film to form a water plasma when the probe beam was above 2 µJ. Then, the 400 nm SH beam emitted from the water plasma was measured by a PMT through a 400 nm narrow-band filter. When a 50 µm thick type-I β-barium borate (BBO) crystal was placed in the path of the 800 nm probe beam, a CSH beam could be generated and collinearly propagated with the fundamental beam. When the CSH beam was vertically polarized and coherent with the TISH beam, a linear component that was positively correlated with the THz field appeared in the collected signal. This provided a method for simultaneously measuring the THz field amplitude and phase. To demonstrate the feasibility of this water-based detection scheme, we measured the THz time-domain waveform and its

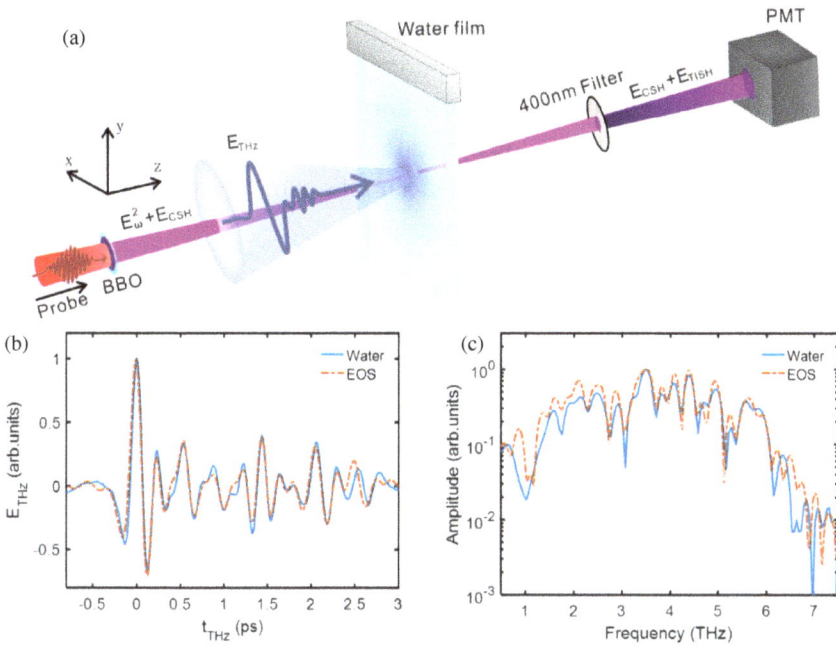

▲ Fig. 10.5. Coherent detection of THz field by water film. (a) Schematic of the experimental setup. An 800 nm laser beam passes through a BBO crystal to generate a CSH of 400 nm and a THz pulse is focused on a water film to generate the TISH. Combining the TISH with CSH, the signal is collected by the PMT. (b), (c) Measured THz waveforms and the corresponding spectra, where the blue solid and red broken lines in each plot correspond to the results of water-based scheme and EOS with a GaP crystal, respectively. Reprinted from Ref. [16] with permission.

corresponding spectrum, as shown by the blue solid lines in Figs. 10.5(b) and (c), respectively. The probe beam energy was set to 5 µJ.

To compare the measured results with electrooptic sampling (EOS), we limited the bandwidth of the THz pulse below 6 THz by using a low-pass filter and controlled the THz field strength at 0.3 MV/cm to ensure the detection validity based on the 100 µm thick GaP crystal. Considering the complex response function, we reconstructed the waveform and spectrum of the THz pulse [17–20], as shown by the red broken lines in Figs. 10.5(b) and (c). The measured results of the THz waveforms and spectra from the two detection methods agreed well, which proved the reliability of

water-based coherent detection. Note that the structured spectrum in Fig. 10.5(c) resulted from the absorption of water vapor rather than liquid water [21]; thus, the characteristic absorption lines of the two methods were almost the same. The introduction of the liquid water film did not obviously change the water vapor concentration; meanwhile, the humidity in our laboratory was maintained constant.

The four-wave mixing process is the basis of air-based THz coherent detection, which is triggered to generate a TISH carrying THz field information. Analogously, four-wave mixing can also be expected to occur in water because it has a higher third-order nonlinear coefficient than air. In the four-wave mixing, the TISH field E_{TISH} depends on the probe laser field E_{ω} and THz field E_{TISH} as: $E_{TISH} = \chi^{(3)} E_{\omega}^2 E_{THz}$, where $\chi^{(3)}$ is the third-order nonlinear coefficient. The TISH energy signal sampled at a time point t_{THz} can be expressed as

$$\varepsilon_{TISH}(t_{THz}) \propto \int_{-\infty}^{+\infty} \left| \chi^{(3)} E_{THz}(t) E_{\omega}^2(t - t_{THz}) \right|^2 dt \qquad (10.8)$$

The TISH energy is proportional to the THz intensity $E_{THz}^2(t_{THz})$ and therefore induces incoherent detection. The TISH energy signal calculated using Eq. (10.8) is shown by the red broken line in Fig. 10.6(a), which reflects the THz intensity distribution. The blue line illustrates the measured signal without the BBO crystal placed in the probe beam (i.e., without the CSH), which approaches the calculated results shown by the red broken line. In this experiment, THz pulses with a strength of 1 MV/cm and frequency range of 0.1–18 THz were used. The inset in Fig. 10.6(a) shows the quadratic dependency of the TISH energy on the THz field strength and its agreement with Eq. (10.8). These results, together with the THz polarization dependence shown below, indicate the attribution of the underlying mechanism to the four-wave mixing.

To realize coherent detection, we introduced a CSH beam by placing a BBO crystal in the optical path of the probe beam. The CSH beam was spatiotemporally overlapped with the TISH beam. For the parallel

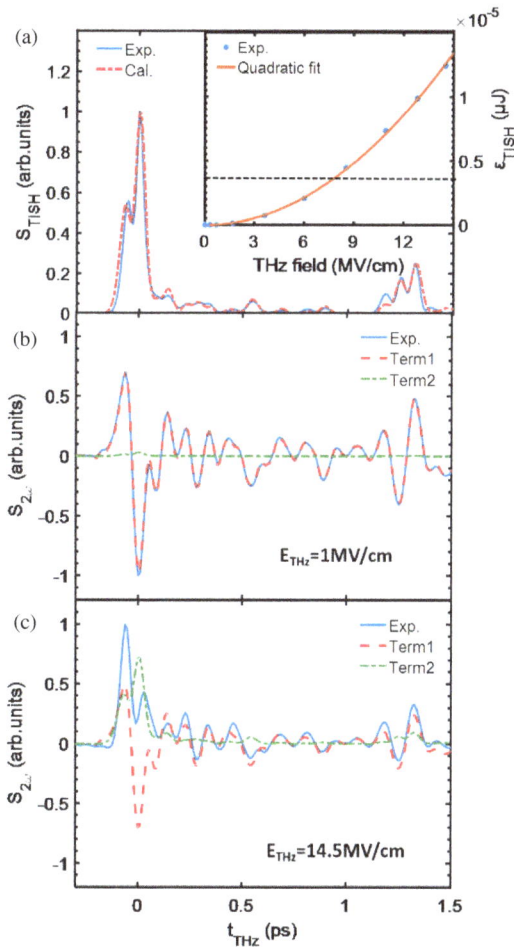

▲ Fig. 10.6.　TISH energy and interference signals from the water film. (a) Measured (blue solid line) and calculated (red broken line) TISH energy signals without the CSH imposed in the detection. Inset shows the TISH energy as a function of the THz field strength, where a quadratic fit is presented by the red solid line. The black broken line in the inset marks the CSH energy used in the experiments of Figs. 10.6(b) and (c). (b), (c) Interference signals of the TISH and CSH, where the THz field strength is considered as 1 MV/cm and 14.5 MV/cm, respectively. The blue solid line shows the measured result, and the red and green broken lines correspond to the calculation results of the first term (coherent component) and second term (incoherent component) in Eq. (10.9), respectively. Reprinted from Ref. [16] with permission.

polarization directions of the CSH and TISH beams, the measured signal at a time point t_{THz} can be expressed as the interference between the two beams:

$$S_{2\omega}(t_{THz}) \propto \int_{-\infty}^{+\infty} |E_{TISH} + E_{CSH}|^2 dt$$

$$\propto 2\,\mathrm{Re}\left\{\int_{-\infty}^{+\infty} \chi^{(3)} E_\omega^2 (t - t_{THz}) E_{CSH}^* (t - t_{THz}) E_{THz}(t) dt\right\} \tag{10.9}$$

$$+ \int_{-\infty}^{+\infty} \left|\chi^{(3)} E_\omega^2 (t - t_{THz}) E_{THz}(t)\right|^2 dt$$

$$+ \int_{-\infty}^{+\infty} E_{CSH}(t - t_{THz}) E_{CSH}^* (t - t_{THz}) dt$$

The first term is linearly proportional to the THz field and provides a phase-resolved cross-correlation measurement that introduces coherent detection. The second term leads to incoherent detection and has been discussed in Eq. (10.8). The third term is a constant background that can be experimentally excluded by lock-in detection. To clarify the contributions of the first two terms, we purposely set the energy of the CSH beam to 3.5×10^{-6} μJ, which was equal to the TISH energy induced by the THz field of 7.5 MV/cm. This is indicated by the black broken line in the inset of Fig. 10.6(a). When the THz field strength was 1 MV/cm and the CSH energy was nearly two orders of magnitude higher than the TISH energy, the coherent component predicted by the first term in Eq. (10.9) was dominant. As shown in Fig. 10.6(b), the measured signal agreed very well with the result calculated using the first term. If we consider the detection with coherent component 10 times higher than the incoherent component as sufficiently coherent, the energy ratio between CSH and TISH should be larger than 25. While the TISH energy is higher, e.g., Fig. 10.6(c) with the THz field strength of 14.5 MV/cm, the incoherent component (green broken line) is at a similar level to the coherent component (red broken line), and hence the measured signal (blue solid line) deviates from the results of the coherent detection.

Note that the maximum SH conversion efficiency of the BBO crystal is ~5% and the CSH energy could reach around 0.25 μJ with the probe beam of 5 μJ. This CSH energy was five orders of magnitude higher than 3.5×10^{-6} μJ used in the experiments shown in Figs. 10.6(b) and (c).

Therefore, it is quite easy to coherently detect strong THz pulses, which requires an energy ratio larger than 25 between CSH and TISH larger. For example, the required CSH energy was only 3.3×10^{-4} µJ to coherently detect a THz field of 14.5 MV/cm, which was the highest field strength available in our experiments. Even when the THz field strength was as high as 100 MV/cm, the required CSH energy of 1.55×10^{-2} µJ was very easy to achieve.

10.3.1 Phase and Polarization Dependencies

We focused on coherent detection for an energy ratio, larger than 25, between CSH and TISH. In this case, the measured time-resolved signal at a time point t_{THz} reduces to

$$S_{2\omega}^{coherent}(t_{THz}) \propto \mathrm{Re}\left\{ \int_{-\infty}^{+\infty} \chi^{(3)} E_{\omega}^2(t-t_{THz}) E_{CSH}^*(t-t_{THz}) E_{THz}(t)dt \right\} \quad (10.10)$$

We defined $E_{\omega}^2(t)E_{CSH}^*(t) = I_{\omega}(t)\sqrt{I_{CSH}(t)}\exp(i\Delta\varphi)$, where the relative phase $\Delta\varphi = 2\varphi_{\omega} - \varphi_{CSH}$ with φ_{ω} and φ_{CSH} denote the phases of the probe and CSH beams, respectively. The signal at time point t_{THz} can be expressed as

$$S_{2\omega}^{coherent}(t_{THz}) \propto \cos(\Delta\varphi)\int_{-\infty}^{+\infty} I_{\omega}(t-t_{THz})\sqrt{I_{CSH}(t-t_{THz})}\,\mathrm{Re}[E_{THz}(t)]dt$$
$$+ \sin(\Delta\varphi)\int_{-\infty}^{+\infty} I_{\omega}(t-t_{THz})\sqrt{I_{CSH}(t-t_{THz})}\,\mathrm{Im}[E_{THz}(t)]dt.$$
$$(10.11)$$

For a given frequency component of the THz field, $\mathrm{Im}[E_{THz}(t)] = \mathrm{Re}[E_{THz}(\omega_{THz}t + \pi/2)]$ and the signal at time point t_{THz} can then be expressed as

$$S_{2\omega}^{coherent}(t_{THz}) \propto \cos(\Delta\varphi)\mathrm{Re}[E_{THz}(t_{THz})] + \sin(\Delta\varphi)\mathrm{Re}\left[E_{THz}(t_{THz})\exp\left(\frac{i\pi}{2}\right)\right]$$
$$(10.12)$$

The measured signal can accurately reproduce the time-resolved THz field during $\Delta\varphi = 0$. Besides, Eq. (10.12) allows us to further investigate the properties of our detection scheme by adjusting the $\Delta\varphi$. To capture the

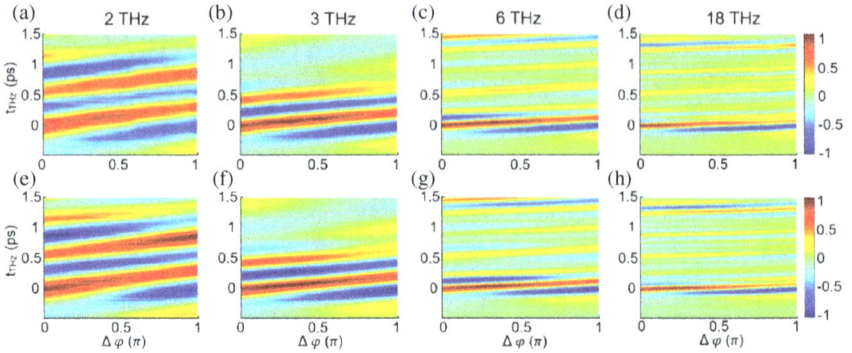

▲ Fig. 10.7. Recorded signals versus time delay and relative phase. (a)–(d) Measured THz signals for the bandwidths of 2, 3, 6, and 18 THz, respectively. (e)–(h) Corresponding calculation results. Reprinted from Ref. [16] with permission.

full dependence of the measured signal on the relative phase, we recorded the THz time-domain waveforms while translating the BBO crystal along the optical path to control the $\Delta\varphi$. Fig. 10.7(a)–(d) illustrate the measured signals as a function of $\Delta\varphi$ and t_{THz} when the THz pulses pass through low-pass filters with cut-off frequencies of 2, 3, 6, and 18 THz, respectively. With the change in $\Delta\varphi$, the measured THz waveforms show a relevant phase shift and flip over when $\Delta\varphi$ changes by π, which is consistent with the simulation results predicted by Eq. (10.12), as shown in Figs. 10.7(e)–(h).

To further demonstrate that the scheme is sensitive to THz wave polarization, we studied the dependency of the TISH energy and interference signals on the relative polarization angle θ between the probe beam and THz field. In the experiment, the probe beam was vertically polarized and the THz polarization was rotated. The TISH energy versus the θ without the CSH beam imposed on the probe beam is shown in Fig. 10.8(a). The vertical component of TISH energy corresponding to $\chi_{yyyy}^{(3)}$ follows $\cos^2\theta$, which agrees with the prediction of Eq. (10.8). The horizontal component of the TISH energy corresponding to $\chi_{xxyy}^{(3)} \approx \frac{1}{3}\chi_{yyyy}^{(3)}$ was also observed, which follows $\sin^2\theta$. Fig. 10.8(b) shows the dependency of the interference signal on θ when the CSH beam was applied to ensure coherent detection and its polarization was vertical. This follows $\cos\theta$ and

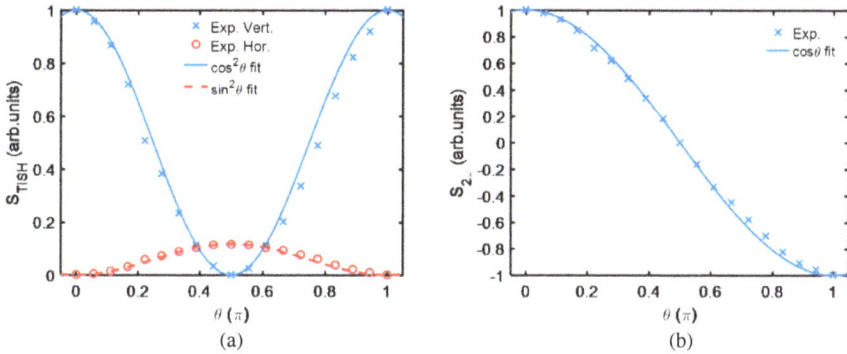

▲ Fig. 10.8.   Dependence of the measured signal on THz polarization. (a) TISH energy as a function of the relative polarization angle between the probe laser and THz fields, where the crosses and dots correspond to the vertical and horizontal components, respectively. (b) Measured signal as a function of cos θ in the case of coherent detection. Reprinted from Ref. [16] with permission.

agrees with the first term in Eq. (10.9). The results indicate that our scheme can provide polarization-sensitive detection and that the two orthogonal components of the THz field can be time resolved by properly rotating the polarization of the probe beam.

10.3.2 Comparison with Air-based Detection

Water-based schemes require much lower probe laser energy than air-based detection schemes [22–25] to generate the same level of TISH. We employed a ~5 µJ probe beam to measure the waveform of a 1 MV/cm THz field in the range of 0.1–18 THz in the water film, as shown by the blue line in Fig. 10.9(a). In the air-based scheme, the energy of the probe beam had to be enhanced to ~75 µJ to coherently detect the THz field (red broken line in Fig. 10.9(a)) at the same level of SNR. A clearer comparison of the two schemes can be seen in Fig. 10.9(b), which shows that probe beam energies 1–2 orders of magnitude higher are needed in air to achieve the same level of TISH energy. In other words, the detection sensitivity in water was one to two orders of magnitude higher than that in air under comparable experimental conditions. For example, if a 35 µJ probe laser

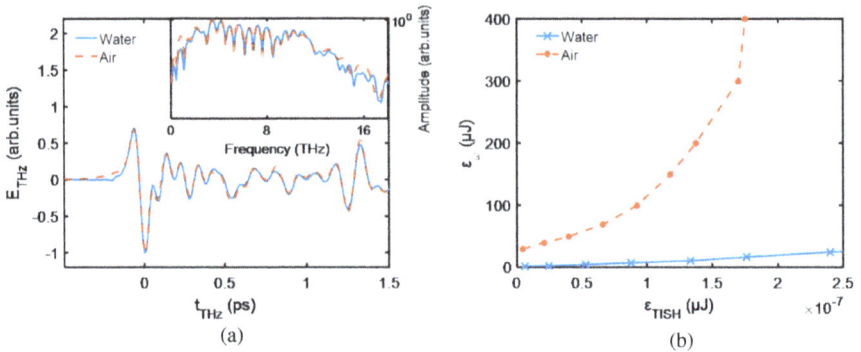

▲ Fig. 10.9.   Comparison of the water- and air-based detection methods. (a) THz time-domain waveforms and the corresponding spectra (inset). (b) Required probe beam energy to generate the TISH energy for the given THz field of 1 MV/cm. In each plot, the blue solid and red broken lines correspond to results with the water- and air-based schemes, respectively. Reprinted from Ref. [16] with permission.

beam was applied, a 0.3 MV/cm THz field could induce 3.1×10^{-8} μJ TISH beam in water, but the value of TISH was only 0.05×10^{-8} μJ in air. With these TISH beams adopted in coherent detection, the SNRs were ~85:1 and ~10:1 for the water- and air-based schemes, respectively. Even when the energy was reduced to 2 μJ, the THz signal could still be observed with a lower SNR.

10.3.3 Comparison of TISH Energy in Water- and Air-based Detection Methods

Fig. 10.10 further demonstrates that lower probe energy was needed to generate the same level of TISH in water than in air. We measured the dependency of the TISH energy on the probe laser energy in air (Fig. 10.10(a)) and water (Fig. 10.10(b)), where in each plot different lines correspond to the THz field strengths of 0.3, 1, and 3 MV/cm. For clarity, the TISH energy is magnified 100 or 10 times with the THz field strength of 0.3 or 1 MV/cm, respectively. Under the same focusing condition in our experiments, the plasma started to get excited from the probe laser energies of 35 and 2 μJ in air and water, respectively. The results of Fig. 10.10 demonstrate that one to two orders of magnitude lower probe

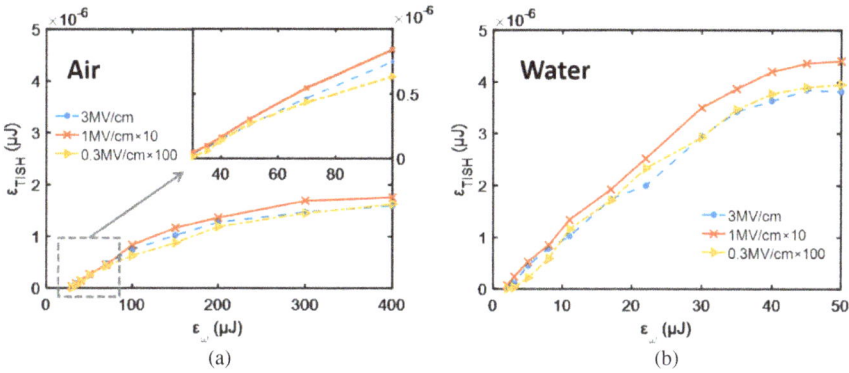

▲ Fig. 10.10.   Dependence of the TISH energy on probe laser energy. Measured TISH energy as a function of the probe laser energy with different THz field strengths of 0.3, 1, and 3 MV/cm. Plots (a), (b) correspond to the air- and water-based detection methods, respectively. For clarity, the TISH energy is magnified 100 or 10 times with the THz field strength of 0.3 or 1 MV/cm, respectively. Reprinted from Ref. [16] with permission.

beam energy is required in water to achieve the same level of TISH energy under comparable experimental conditions. For instance, to generate the TISH energy of $1 \times 10^{-6}\,\mu J$ with a 3 MV/cm THz field, the probe laser energy of $\sim 10\,\mu J$ was required in water, which was one order of magnitude lower than the energy of $\sim 150\,\mu J$ required in air. It can be seen from Fig. 10.10 that under the same experimental conditions (considering the detection light with a single pulse energy of 50 μJ as an example), the energy of the SH induced by THz waves in liquid water is approximately 20 times that in air plasma. At this time, for a THz pulse with an electric field intensity of ~ 300 kV/cm, the conversion efficiency of SH is approximately 8×10^{-10} (approximately 4×10^{-11} in air).

10.4 Detection of Broadband THz Wave using Aqueous Salt Solutions

Here, we demonstrated that the water-based detection method can be extended to aqueous salt solutions and that the sensitivity can be significantly enhanced owing to the typically higher third-order nonlinear coefficient $\chi^{(3)}$ of aqueous salt solutions [26, 27]. Meanwhile, the scheme provides the possibility of revealing the $\chi^{(3)}$ or refractive index of various liquids in the THz range. The physicochemical properties of different salt solutions vary significantly [28–31]. The properties of pure water in the THz band have been studied by THz spectroscopy [32, 33], ultrafast infrared spectroscopy [34, 35], and THz Kerr effect spectroscopy [19, 36–38]. However, the physicochemical properties of aqueous salt solutions in the THz band have not been systematically studied. In addition, previous studies using spectroscopic techniques have both advantages and disadvantages, whereas liquid-based THz coherent detection provides valuable complementary information of aqueous salt solutions.

10.4.1 Enhanced Detection Sensitivity with Solutions

Broadband THz pulse coherent detection experiments were performed using different salt solutions [39]. Fig. 10.11(a) shows the time-domain waveforms of the THz pulses detected by aqueous iodide solutions (CsI, LiI, NaI, and KI). The detection sensitivity of the solution-based scheme is higher than that of pure water. All experiments were performed under the same conditions. The signal enhancement was attributed to the addition of salt ions. The THz signal amplitude was significantly enhanced in the order CsI > LiI > NaI > KI > water. The concentration of all salt

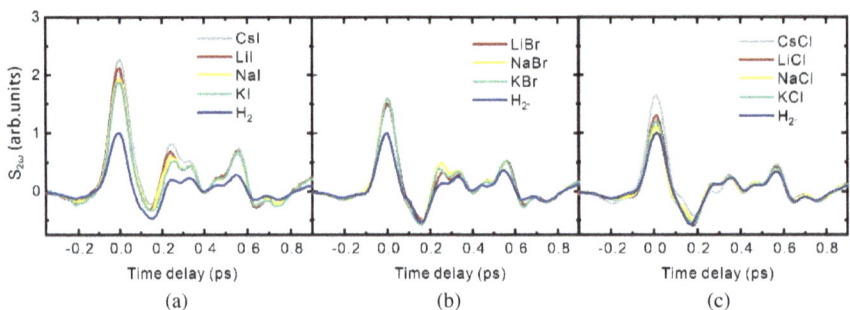

▲ Fig. 10.11.   Coherent detection of THz field in different solution films. THz time-domain waveforms detected in (a) iodide aqueous solution (LiI, NaI, KI, CsI), (b) bromide aqueous solution (LiBr, NaBr, KBr), and (c) chloride aqueous solution (LiCl, NaCl, KCl, CsCl). THz waveforms detected by pure water are displayed in each diagram for comparison. Reprinted from Ref. [39] with permission.

solutions was 4 mol/L, except for CsI, since CsI reaches saturation when the concentration is over 2 mol/L. In particular, the detection sensitivity of CsI increased by ~2.3 times. Although the concentration of CsI was the lowest, the enhancement in detection sensitivity was the strongest. The results revealed that the detection sensitivity of the iodide aqueous solution was higher than that of bromide and chloride aqueous solutions at the same concentration. Although the detection sensitivity of cesium ions was still the highest, it was not arranged in ascending order of the ionic atomic weight. The results are presented in Figs. 10.11(b) and (c). CsBr was not included in this experiment because of its high toxicity and cost. The results showed higher sensitivity of the salt solution than that of pure water.

The refractive index η_0 of the solution is proportional to the refractive index η_s of the solid salt and its concentration, as shown in Fig. 10.12(a). This is demonstrated in Eqs. (10.13) and (10.14), respectively, in Section 10.4.2. Here, we considered the refractive index η_s of solid salts

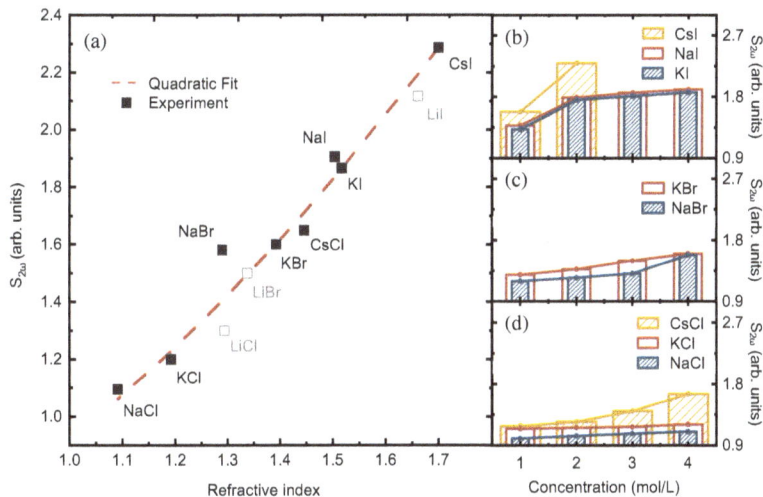

▲ Fig. 10.12. Coherent detection signals from different solutions. (a) Measured signal intensity as a function of linear refractive index (black/white squares), where the red dashed line represents the quadratic fitting curve. The measured signal intensity depends on the concentration of (b) aqueous iodide solution, (c) aqueous bromide solution, and (d) aqueous chloride solution. Reprinted from Ref. [39] with permission.

at a frequency of 9 THz. In Figs. 10.12(a) and 10.13(a), we still matched the experimental results as listed refractive indices of the three types of salt, which are distinguished by the hollow selvedge square. It is clear that the detection signal intensity increases monotonically with the η_0 of different solutions, and quadratically depends on the η_0 of different solutions. Fig. 10.12(b)–(d) illustrate that the signal intensity increases with increasing solution concentration. This strength enhancement was attributed to the increase in the refractive index with increasing solution concentration.

10.4.2 Scaling Law of Signal Intensity of Coherent Detection

By comparing the square representing the experimental results with the quadratic fitting curve in Fig. 10.12(a), we perform a rough scaling of the signal intensity and refractive index η_0 of the salt solution in a quadratic curve. The quadratic scaling is attributed to four-wave mixing because the signal $S_{2\omega}$ linearly depends on the third-order nonlinear coefficients $\chi^{(3)}$, which are quadratically proportional to the linear refractive index η_0, as given by

$$S_{2\omega} \propto \chi^{(3)} \propto \eta_0^2. \tag{10.13}$$

With the increase in solution concentration C, the relationship between the $\chi^{(3)}$, η_0, and η_s can be obtained as

$$\chi^{(3)} \propto \eta_0^2 \propto C\eta_s^2. \tag{10.14}$$

Eq. (10.14) explains the signal intensity scaling, as shown in Fig. 10.12(a). Fig. 10.12(b)–(d) show that the signal intensity of coherent detection increases with an increase in the solution concentration, and the slope of the signal intensity also changes. In Figs. 10.12(b)–(d), the salt with a higher refractive index has a higher signal amplitude, which is in agreement with the result in Fig. 10.12(a). It should be noted that for the NaI and KI solutions in Figs. 10.11(b)–(d), the linear increase in the signal amplitude remains unchanged at lower concentrations. However, the rate of increase slows down at higher concentrations as the saturation approaches.

The second term of Eq. (10.9) corresponds to incoherent detection, where the relation between the coherent and incoherent components as in $|\chi^{(3)}E_\omega^2 E_{THz}|^2 \ll |\chi^{(3)}E_\omega^2 E_{CSH}^* E_{THz}|$ is satisfied. The signal contained only TISH beams. Fig. 10.13(a)–(c) show the TISH signals measured in the aqueous iodide, bromide, and chloride solutions, respectively. By combining Eq. (10.8) and (4.1.20) in Ref. [10], we obtain:

$$S_{TISH} \propto [\chi^{(3)}]^2 \propto \eta_0^4 \propto C^2 \eta_S^4, \tag{10.15}$$

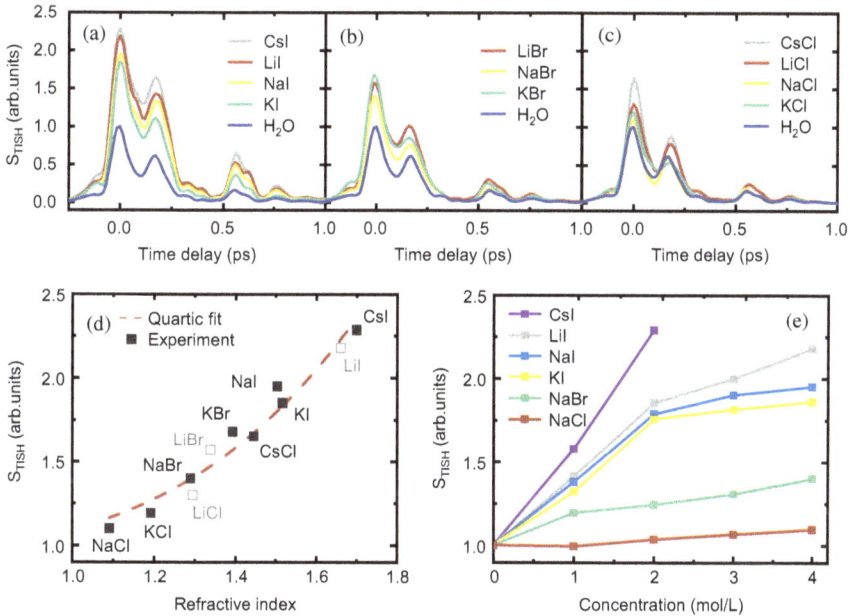

▲ Fig. 10.13.   Incoherent detection in different salt solutions. The TISH signal detected in (a) aqueous iodide solution (LiI, NaI, KI, CsI), (b) aqueous bromide solution (LiBr, NaBr, KBr), (c) aqueous chloride solution (LiCl, NaCl, KCl, CsCl), and the TISH signal detected in pure water are used for comparison. (d) The dependence of measured TISH energy on the concentration of six representative solutions. (e) The inset shows the measured TISH energy as a function of the linear refractive index of different solutions, where the four-fold fitted curves are represented by the red dashed lines. Reprinted from Ref. [39] with permission.

which explains the approximate quartic fit of the incoherent signal or S_{TISH} with the refractive index of the 11 aqueous solutions observed in our experiments, as shown in Fig. 10.13(d). Figure 10.13(e) shows that the signal intensity increases with an increase in the solution concentration, which is similar to the results shown in Figs. 10.11(b)–(d). A comparison of the slopes of the signal intensity shows that the variation in the incoherent signal in Fig. 10.13(e) is faster than that of the coherent signal in Figs. 10.12(b) and (d), which is in good agreement with Eqs. (10.14) and (10.15).

10.4.3 THz Time-Domain Waveform Comparison in Coherent Detection

The normalized THz time-domain waveforms detected by coherent detection in different aqueous salt solutions are shown in Figs. 10.14(a)–(c). Compared with pure water (blue curve), the waveform of the first half-period maintains the same shape, while that of the second half-period (or the first negative half-period) appears to be attenuated in the aqueous salt solutions. We calculated the differences in the THz field amplitude in the second half-period between the aqueous salt solutions and pure water. The corresponding dependence of the linear refractive index of the solid salt in solution is shown in Fig. 10.14(d). The results show that attenuation is positively correlated with the refractive index. Note that the TISH field was generated when liquid plasmas were formed in our experiment. The ionization excited by the probe laser beam mainly occurred in the first half of the THz pulse because the laser duration was approximately 50 fs. The THz pulse and formed plasma mainly interacted after the first half of the THz pulse, resulting in significant attenuation in the second half of the THz time-domain waveform. In addition, in salt solutions with a higher refractive index (higher third-order nonlinear susceptibility), the laser beam tended to form stronger plasma filaments, resulting in a greater attenuation of the transmitted THz field.

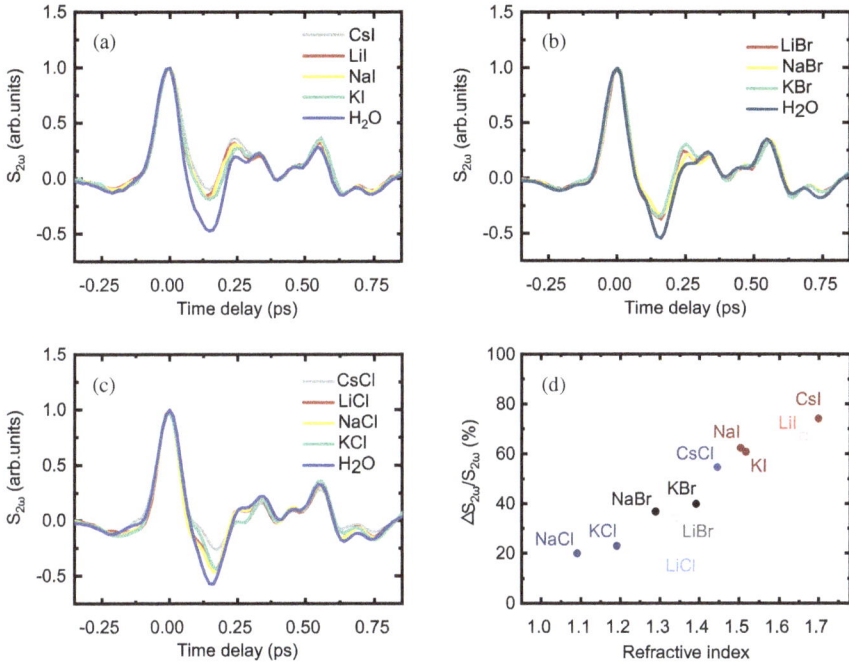

▲ Fig. 10.14. Normalized THz time-domain waveform detected by (a) aqueous iodide solutions (LiI, NaI, KI, CsI), (b) aqueous bromide solutions (LiBr, NaBr, KBr), and (c) aqueous chloride solutions (LiCl, NaCl, KCl, CsCl). The concentration of all aqueous salt solutions was 4 mol/L except that of CsI solution was set as 2 mol/L. The waveforms detected by pure water are displayed in each diagram for comparison. (d) Differences of the amplitude in the second half-cycle between aqueous salt solutions and pure water. Reprinted from Ref. [39] with permission.

10.5 Efficient Coherent Detection of Terahertz Pulses Based on Ethanol

Despite the great potential of using liquid water for THz wave generation and detection [14, 16, 40, 41], the response of water to THz waves is weaker than that of other liquids, such as ethanol [42]. Ethanol has a lower ionization energy than water [43, 44], so it is more easily ionized and requires a lower probe laser energy to form liquid plasma. Moreover, the third-order nonlinear coefficient of ethanol is larger than that of water

[45, 46]. Therefore, it is reasonable to expect a more promising application of ethanol in THz aqueous photonics. Our previous work confirmed that ethanol has a higher molecular response than pure water in the THz band [10, 47]. Furthermore, studies on THz wave generation from liquids have verified that ethanol can emit stronger THz waves than pure water [19]. Generally, the generation process can be reversed for THz wave detection. The liquid water used for coherent detection of THz waves was experimentally and theoretically confirmed. However, a THz wave coherent detection scheme using ethanol or ethanol-water mixtures has not yet been investigated.

Here, we demonstrated that ethanol is an efficient THz wave coherent detection medium [48]. The underlying mechanism is attributed to four-wave mixing. The waveform of the THz pulse was obtained by combining TISH and CSH beams. Coherent, incoherent, and hybrid detections could be alternately realized by rotating the polarization of the CSH. Our scheme was well implemented with a considerably lower probe laser energy of even a few microjoules. The experimental results showed further enhanced sensitivity and SNR of coherent detection using ethanol compared with pure water. In addition, we performed THz coherent detection using ethanol-water mixtures with different concentrations and revealed that the contribution of the mixture can always be decomposed into the synergistic effect of ethanol and water molecules within the sub-picosecond timescale.

A schematic of our experimental setup is shown in Fig. 10.15(a). A vertically polarized intense broadband THz source was generated by an organic DAST crystal under the excitation of a femtosecond laser pulse with a central wavelength of 1550 nm and a repetition rate of 1 kHz. The THz beam was focused onto the liquid film using a two-inch off-axis parabolic mirror to excite the TISH beam. The energy of the THz field was controlled using two wire-grid THz polarizers. The residual optical beam was filtered using a set of low-pass filters with cut-off frequency of 9 THz. The 800 nm central wavelength probe beam was collinearly propagated with the THz beam and was focused onto the liquid film by a four-inch

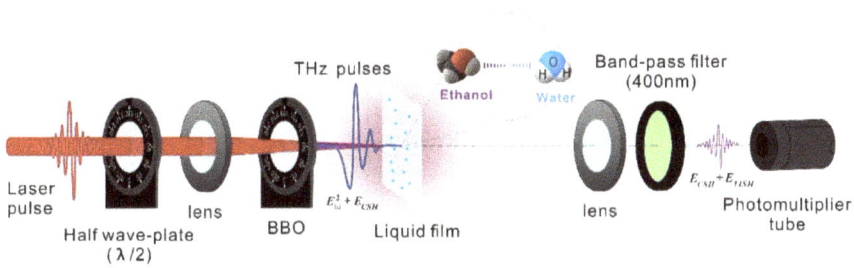

▲ Fig. 10.15.   Experimental system diagram. An 800 nm laser beam passes through a BBO crystal to generate a CSH of 400 nm and a THz pulse is focused on a water film to generate the TISH. Combining the TISH with CSH, the signal is collected by the PMT. Reprinted from Ref. [48] with permission.

quartz lens to excite the plasma. A half-wave plate (HWP) was used to rotate the polarization of the probe beam. A 100 μm thick BBO crystal was appropriately placed in the optical path to generate a CSH beam. Subsequently, the intensity of the SH beam was observed using a PMT through a 400 nm narrow band-pass filter.

10.5.1 Comparison of Signals Measured by Ethanol, Water and Air

In our experiment, the peak THz field strength generated by the DAST crystal was 4.6 MV/cm. The probe laser energy was set at 15 μJ. Because the maximal SH conversion efficiency in the BBO crystal was approximately 5%, the maximum energy produced by the CSH beam was approximately 0.75 μJ. Although, in this case, the polarization of the CSH beam might not be parallel to the THz field to ensure fully coherent interference between the CSH and TISH beams (which will be discussed in Section 10.5.2), it could be rotated by inserting a 400 nm half-wave plate after the BBO crystal. Therefore, the requirements for coherent detection could be easily satisfied in this experiment. The THz time-domain waveforms detected by ethanol, water, and air under the same THz field strength and probe laser energy are depicted in Fig. 10.16(a). A significant enhancement of the THz coherent detection signal amplitude in the order of ethanol > water > air was observed.

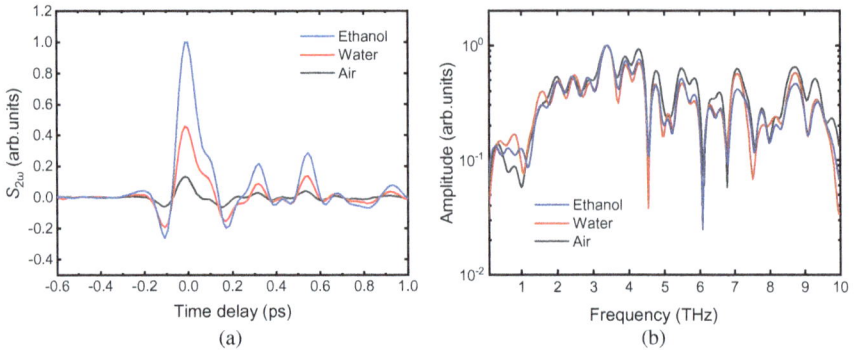

▲ Fig. 10.16.   Coherent detection of THz filed by liquid film. (a) Measured THz wave-forms detected by ethanol (blue line), pure water (red line), and air (black line). (b) The corresponding frequency spectra. Reprinted from Ref. [48] with permission.

A comparison between water and air has been discussed in Ref. [16]. For ethanol and water, as shown in Eq. (10.9), the third-order nonlinear coefficient $\chi^{(3)}$ of the liquid plays an important role in coherent detection. The values of the linear refractive index coefficient n_0 and nonlinear refractive index coefficient n_2 of ethanol and water in the THz spectral region were calculated and measured [49]. The ratios of n_0 and n_2 between ethanol and water were approximately 1:1.5 and 8.6:1, respectively. Therefore, we can conclude that the $\chi^{(3)}$ of ethanol is ~3.8 times larger than that of water according to Eq. (4.1.12) in Ref. [10]:

$$\chi^{(3)}\left(\frac{m^2}{V^2}\right) = \frac{n_0^2}{283} n_2 \left(\frac{m^2}{W}\right) \tag{10.16}$$

Theoretically, the $S_{2\omega}$ of ethanol is larger than that of both water and air, which agrees well with our experimental results. Figure 10.16(b) shows that the corresponding frequency spectra measured by the three different media are in good agreement. Note the structured spectrum results from the absorption of water vapor; thus, the characteristic absorption lines of the three spectra are almost the same, owing to the constant humidity in our laboratory.

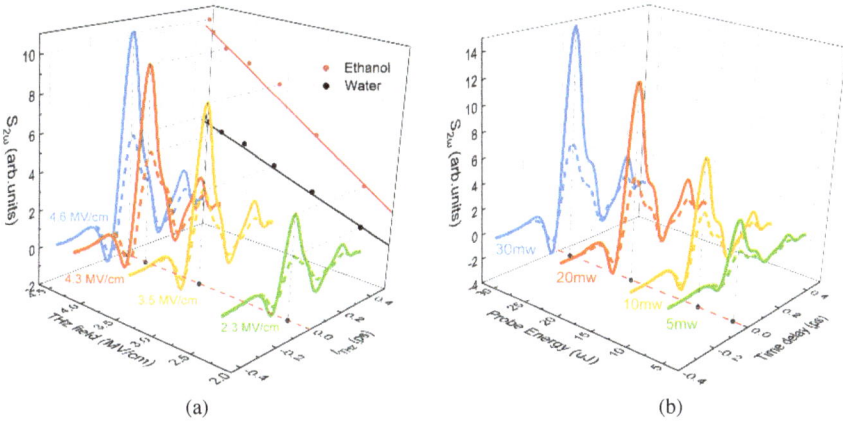

(a)   (b)

▲ Fig. 10.17.   (a) THz time-domain waveforms measured at different THz electric field strength detected by ethanol (solid lines) and water (dotted lines). The linear fitting results represent the ethanol (red dots) and water (black dots) detection signal peaks to THz electric field intensities. (b) Waveforms detected by ethanol (solid lines) and pure water (dotted lines) under different probe laser energy from 5 μJ to 30 μJ, respectively. (a) is reprinted from Ref. [48] with permission.

To demonstrate that the coherent component $S_{2\omega}^{coherent}$ described in the first term of Eq. (10.9) is linearly proportional to the THz electric field intensity E_{THz}, when the probe beam power was fixed at 15 μJ, the measurement results under different THz electric fields were as shown in Fig. 10.17(a). The fitted linear line was in good agreement with the peak amplitude detected by THz wave excitation in ethanol and pure water. Moreover, the slope of the fitted line indicates more sensitivity of ethanol than water in the THz band. We compared the coherent detection signals of ethanol and pure water at different detection light energies ranging from 5 μJ to 30 μJ. The THz electric-field intensity was fixed at 3.5 MV/cm to ensure sufficient coherent detection when the CSH beam energy was low. The measurement results are shown in Fig. 10.17(b). Clearly, ethanol exhibits a higher measurement value than pure water for any detectable light energy. In addition, the time-domain waveform of ethanol still had a good SNR, even when the detection beam energy was as low as 5 μJ, which provides a new research perspective for THz coherent detection at low laser energy.

10.5.2 Sensitive Detection of CSH Beam Polarization

To further investigate the effect of relative polarization of the CSH on coherent detection, a space coordinate system was established. α was defined as the azimuthal angle between the $\hat{e}$-axis of the BBO crystal and horizontal direction (x-axis). The CSH wave generated from the BBO was polarized along its $\hat{e}$-axis. The slight change in the polarization direction of the fundamental wave by rotating angle α was negligible. The projection of the CSH field along the y-axis is given by:

$$E^y_{CSH} = E_{2\omega} \sin\alpha$$
$$E_{2\omega} = d_{eff} E^2_\omega \cos^2\alpha, \tag{10.17}$$

where E_ω and $E_{2\omega}$ are the amplitudes of the fundamental (ω) and SH (2ω) fields, respectively, and d_{eff} is the effective conversion coefficient of the BBO crystal [50]. The polarization of the THz field was parallel to the y-axis, introducing a vertically polarized TISH beam: $E^y_{TISH} = \chi^{(3)}_{yyyy}(E^y_\omega)^2 E^y_{THz}$.

The time-domain waveforms recorded by rotating the BBO azimuth angle are shown in Fig. 10.18. As shown by the gray dashed line, the CSH intensity was maximum at $\alpha = 0°(180°)$ and minimum at $\alpha = 90°(270°)$ [51]. Because the polarization of CSH was always along the $\hat{e}$-axis of the BBO crystal, CSH was horizontally polarized when $\alpha = 0°(180°)$, and the CSH and TISH fields were perpendicular to each other. In this case, interference between the CSH and TISH was impossible, resulting in incoherent detection. Theoretically, the CSH intensity should have been zero during $\alpha = 90°(270°)$, but a detectable CSH signal was still obtained in our experiment (shown by the black dotted line in Fig. 10.18), which could be the result of slight deviation of the normal incident of the probe laser beam on the BBO crystal. The measured vertically polarized CSH beam, parallel to the TISH beam, was of the order of microjoules, satisfying the requirement of the CSH intensity for sufficient coherent detection. Therefore, the first term in Eq. (10.9) dominates. Except for the four angles for sufficiently coherent or incoherent detection, the remaining angles, that

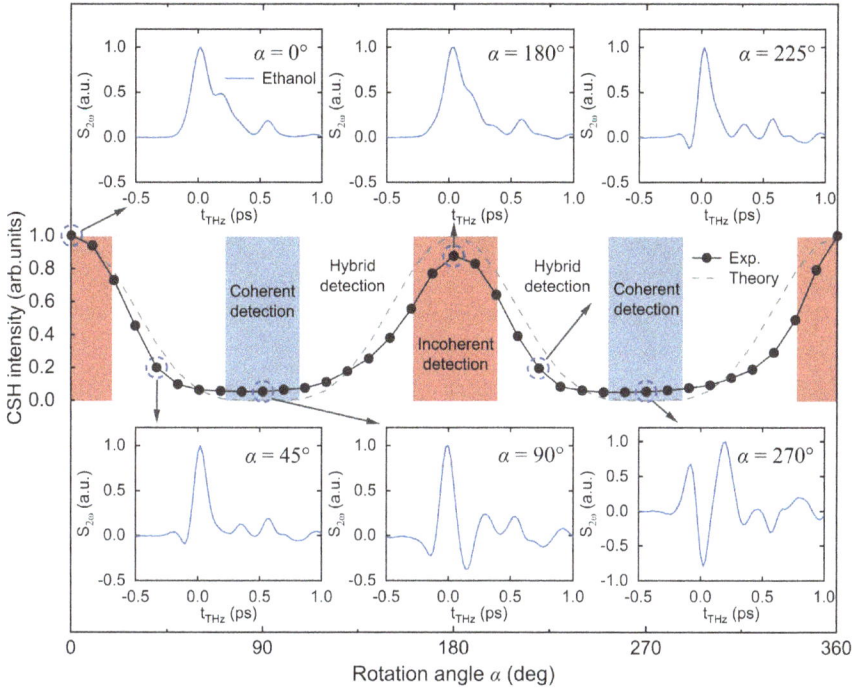

▲ Fig. 10.18. Measured waveforms at different angle α. When $\alpha = 0°(180°)$, CSH is perpendicular polarized to TISH and ω, leading to incoherent detection. When $\alpha = 90°(270°)$, the polarizations of CSH, TISH, and ω are parallel, which presents as coherent detection. At $\alpha = 45°(225°)$ a hybrid detection is introduced. The black dots and gray dashed line are the measured and theoretical CSH energy as a function of angle α, respectively. Reprinted from Ref. [48] with permission.

is, $\alpha = 45°(225°)$, introduced as the first and second terms in Eq. (10.9), were comparable.

10.5.3 THz Coherent Detection using Ethanol-Water Mixtures

In addition, based on the measurement results and theoretical analysis of the coherent detection of pure water and ethanol, we further investigated the THz signals detected by ethanol-water mixtures with different concentrations excited by a THz field strength of 4.6 MV/cm.

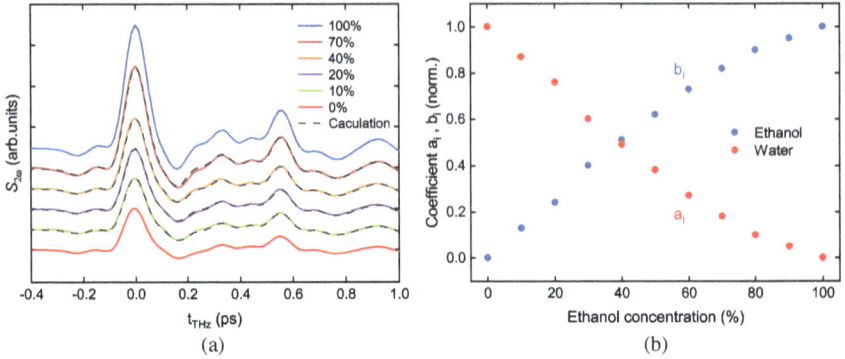

▲ Fig. 10.19. Coherent detection using ethanol-water mixture. (a) Measured THz time-domain waveforms and calculation results (gray dashed lines) of coherent detection by ethanol-water mixtures with different ethanol concentrations. (b) Coefficients a_i, b_i of water and ethanol for different ethanol concentrations. Reprinted from Ref. [48] with permission.

The measured results are depicted in Fig. 10.19(a), which shows an increase in the signal amplitude of the ethanol-water mixtures with increase in the ethanol concentration. The fitted results for the mixtures calculated using Eq. (10.9) are shown by dashed lines in Fig. 10.19(a).

$$\Delta S_{2\omega}(t_{THz}) = a_i S_{2\omega}^{\text{water}}(t_{THz}) + b_i S_{2\omega}^{\text{ethanol}}(t_{THz}), \qquad (10.18)$$

where $S_{2\omega}^{\text{water}}(t_{THz})$ and $S_{2\omega}^{\text{ethanol}}(t_{THz})$ represent the measured coherent detection signals for pure water and ethanol, respectively. As shown in Fig. 10.19(b), a_i and b_i represent the proportional coefficients related to the concentration of ethanol in the mixture. The calculated results are in good agreement with the measured results, indicating a possible division of the temporal evolution curves of the coherent detection of the mixture into linear superposition of the coherent detection signals of pure water and neat ethanol. Apparently, the contribution to coherent detection depends on the number of molecules of ethanol and water at sub-picosecond timescales, revealing the independence of intermolecular vibrational modes in the mixture. Notably, the linear superposition results deviated from the measured data when the timescale was extended beyond a few picoseconds. Owing to the complex interactions related to the molecular structure of the mixture, more in-depth experiments are required for further investigation.

10.6 Conclusion

Our investigation demonstrated that liquid can be used for coherent THz pulse detection. We have achieved the time-domain waveform of the THz field with a frequency range of 0.1–18 THz, where the cut-off frequency of 18 THz was limited by the adopted THz pulse source and low-pass filter available in our experiments. Our scheme could be efficiently realized with a low probe laser energy of approximately 5 μJ. The sensitivity of our scheme was one order of magnitude higher than that of the air-based scheme under comparable detection conditions. The lower energy requirement is because liquid water has a higher third-order nonlinear coefficient and a lower ionization threshold than air, where plasma excitation is necessary in both water- and air-based schemes. Our scheme is also sensitive to THz polarization and the phase difference between fundamental and CSH beams, which introduces flexible methods for optimization and polarization-sensitive detection. We thus hope that we have offered a new outlook to coherently detect broadband THz pulses, which has great potential for THz-TDS applications and THz remote sensing.

We further investigated aqueous salt solution- and ethanol-based THz coherent detection. The measured results revealed that the enhancement of the detected THz field amplitude mainly contributed to the third-order nonlinear coefficient $\chi^{(3)}$ of the liquids. This work provides a new perspective for liquid-based coherent detection and a new research idea for exploring solute-solvent molecular interactions.

References

1. Xu K., Liu M. & Arbab M. H. (2022). Broadband terahertz time-domain polarimetry based on air plasma filament emissions

and spinning electro-optic sampling in GaP, Applied Physics Letters, 120(18), p. 181107.

2. Nahata A., Auston D., Heinz T. & Wu C. (1996). Coherent detection of freely propagating terahertz radiation by electro-optic sampling, Applied Physics Letters, 68(2), pp. 150–152.

3. Wu Q. & Zhang X.-C. (1995). Free-space electro-optic sampling of terahertz beams, Applied Physics Letters, 67, p. 3523.

4. Lu X., Karpowicz N., Chen Y. & Zhang X.-C. (2008). Systematic study of broadband terahertz gas sensor, Applied Physics Letters, 93, p. 261106.

5. Dai J., Xu X. & Zhang X.-C. (2006). Detection of broadband terahertz waves with a laser-induced plasma in gases, Physical Review Letters, 97, p. 103903.

6. Liu J., Dai J., Lu X., Ho I. C. & Zhang X.-C. (2012). Broadband terahertz wave generation, detection and coherent control using terahertz gas photonics, International Journal of High Speed Electronics and Systems, 20, p. 1100635.

7. Buccheri F., Huang P. & Zhang X.-C. (2018). Generation and detection of pulsed terahertz waves in gas: from elongated plasmas to microplasmas, Frontiers of Optoelectronics, 11(3), pp. 209–244.

8. Vugmeyster I. D., Whitaker J. F. & Merlin R. (2012). GaP based terahertz time-domain spectrometer optimized for the 5–8 THz range, Applied Physics Letters, 101(18), p. 181101.

9. Wu B., Cao L., Zhang Z., Fu Q. & Xiong Y. (2018). Terahertz electro-optic sampling in thick ZnTe crystals below the reststrahlen band with a broadband femtosecond laser, IEEE Transactions on Terahertz Science and Technology, 8(3), pp. 305–311.

10. Boyd R. W. (2003). *Nonlinear Optics*. Elsevier.

11. Minardi S., Gopal A., Tatarakis M., Couairon A., Tamošauskas G., Piskarskas R., Dubietis A. & Trapani P. D. (2008). Time-resolved refractive index and absorption mapping of light-plasma filaments in water, Optics Letters, 33, p. 86.

12. Noack J. & Vogel A. (1999). Laser-induced plasma formation in water at nanosecond to femtosecond time scales: calculation of thresholds, absorption coefficients, and energy density, IEEE Journal of Quantum Electronics, 35, p. 1156.

13. Dey I., Jana K., Fedorov V. Y., *et al.* (2017). Highly efficient broadband terahertz generation from ultrashort laser filamentation in liquids, Nature Communications, 8(1), p. 1184.
14. Jin Q., E. Y. & Williams K. (2017). Observation of broadband terahertz wave generation from liquid water, Applied Physics Letters, 111(7), p. 071103.
15. Kim K.Y. (2009). Generation of coherent terahertz radiation in ultra-fast laser-gas interactions, Physics of Plasmas, 16(5), p. 056706.
16. Tan Y., Zhao H., Wang W.-M., Zhang R., Zhao Y.-J., Zhang C.-L., Zhang X.-C. & Zhang L.-L. (2022). Water-based coherent detection of broadband terahertz pulses, Physical Review Letters, 128, p. 093902.
17. Wu Q. & Zhang X.-C. (1997). 7 terahertz broadband GaP electrooptic sensor, Applied Physics Letters, 70, p. 1784.
18. Leitenstorfer A., Hunsche S., Shah J., Nuss M. C. & Knox W. H. (1999). Detectors and sources for ultrabroadband electrooptic sampling: experiment and theory, Applied Physics Letters, 74, p. 1516.
19. Casalbuoni S., Schlarb H., Schmidt B., Schmuser P., Steffen B. & Winter A. (2008). Numerical studies on the electrooptic detection of femtosecond electron bunches, Physical Review Accelerators & Beams, 11, p. 072802.
20. Zhao H., Tan Y., Zhang L. L., Zhang R., Zhang C. L., Zhao Y. J. & Zhang X.-C. (2020). Ultrafast hydrogen-bond dynamics of liquid water revealed by terahertz-induced transient birefringence, Light: Science & Applications, 9, p. 136.
21. Wang T., Klarskov P. & Jepsen P. U. (2014). Ultrabroadband THz time-domain spectroscopy of a free-flowing water film, IEEE Transactions on Terahertz Science and Technology, 4, p. 425.
22. Woerner M. & Reimann K. (2009). Harnessing terahertz polarization, Nature Photonics, 3, p. 495.
23. Liu J., Dai J., Chin S. L. & Zhang X.-C. (2010). Broadband terahertz wave remote sensing using coherent manipulation of fluorescence from asymmetrically ionized gases, Nature Photonics, 4, p. 1.
24. Karpowicz N., Dai J., Lu X., Chen Y., Yamaguchi M., Zhao H. & Zhang X.-C. (2008). Coherent heterodyne time-domain spectrometry covering the entire "terahertz gap", Applied Physics Letters, 92, p. 011131.

25. Chia-Yeh L., Denis V. S., Zhou Y. & Mansoor S. B. (2015). Broadband field-resolved terahertz detection via laser induced air plasma with controlled optical bias, Optics Express, 23, p. 11436.
26. Débarre D. & Beaurepaire E. (2007). Quantitative characterization of biological liquids for third-harmonic generation microscopy, Biophysical Journal, 92, pp. 603–612.
27. Kucia W. E., Sharma G., Joseph C. S., Sarbak S., Oliver C., Dobek A, & Giles R. H. (2016). Optical Kerr effect of tRNA solution induced by femtosecond laser pulses, Chemical Physics Letters, 662, pp. 132–136.
28. Kaminsky M. (1957). Ion-solvent interaction and the viscosity of strong-electrolyte solutions, Discussions of the Faraday Society, 24, pp. 171–179.
29. McAllister R. A. (1960). The viscosity of liquid mixtures, AIChE Journal, 6, pp. 427–431.
30. Lazzús J. A. (2012). A group contribution method to predict the melting point of ionic liquids, Fluid Phase Equilibria, 313, pp. 1–6.
31. Ryckaert J.-P. & Bellemans A. (1975). Molecular dynamics of liquid n-butane near its boiling point, Chemical Physics Letters, 30, pp. 123–125.
32. Funkner S., Niehues G., Schmidt D. A., Heyden M., Schwaab G., Callahan K. M., Tobias D. J. & Havenith, M. (2012). Watching the low-frequency motions in aqueous salt solutions: the terahertz vibrational signatures of hydrated ions, Journal of The American Chemical Society, 134, pp. 1030–1035.
33. Sebastiani F., Wolf S. L. P., Born B., Luong T. Q., Cölfen H., Gebauer D. & Havenith M. (2017). Water dynamics from THz spectroscopy reveal the locus of a liquid–liquid binodal limit in aqueous $CaCO_3$ solutions, Angewandte Chemie International Edition, 56, pp. 490–495.
34. Ohtaki H. & Radnai T. (1993). Structure and dynamics of hydrated ions, Chemical Reviews, 93, pp. 1157–1204.
35. Dodo T., Sugawa M., Nonaka E., Honda H. & Ikawa S. (1995). Absorption of far-infrared radiation by alkali halide aqueous solutions, The Journal of Chemical Physics, 102, pp. 6208–6211.
36. Freysz E. & Degert J. (2010). Terahertz Kerr effect, Nature Photonics, 4, pp. 131–132.

37. Zhao H., Tan Y., Zhang R., Zhao Y., Zhang C. & Zhang L. (2021). Anion-water hydrogen bond vibration revealed by the terahertz Kerr effect, Optics Letters, 46, p. 230.

38. Tan Y., Zhao H., Zhang R., Zhang C., Zhao Y. & Zhang L. (2020). Ultrafast optical pulse polarization modulation based on the terahertz-induced Kerr effect in low-density polyethylene, Optics Express, 28, p. 35330.

39. Zhang M.-H., Xiao W., Wang W.-M., Zhang R., Zhang C.-L., Zhang X.-C. & Zhang L.-L. (2022). Highly sensitive detection of broadband terahertz waves using aqueous salt solutions, Optics Express, 30, pp. 39142–39151.

40. E Y., Jin Q., Tcypkin A. & Zhang X.-C. (2018). Terahertz wave generation from liquid water films via laser-induced breakdown, Applied Physics Letters, 113, p. 181103.

41. Zhang L.-L., Wang W.-M., Wu T., Feng S.-J., Kang K., Zhang C.-L., Zhang Y., Li Y.-T., Sheng Z.-M. & Zhang X.-C. (2019). Strong terahertz radiation from a liquid-water line, Physical Review Applied, 12, p. 014005.

42. Zalden P., Song L. & Wu X. (2018). Molecular polarizability anisotropy of liquid water revealed by terahertz-induced transient orientation, Nature Communication, 9, p. 2142.

43. Watanabe K., Nakayama T. & Mottl T. J. (1962). Ionization potentials of some molecules, Journal of Quantitative Spectroscopy & Radiative Transfer, 2(4), pp. 369–382.

44. Beckey H. D., Levsen K., Röllgen F. W. & Schulten H.-R. (1978). Field ionization mass spectrometry of organic compounds, Surface Science, 70(1), pp. 325–362.

45. Gaiduk A. P., Pham T. A. & Govoni M. (2018). Electron affinity of liquid water, Nature Communications, 9(1), 247.

46. Noack J. & Vogel A. (1999). Laser-induced plasma formation in water at nanosecond to femtosecond time scales: Calculation of thresholds, absorption coefficients, and energy density, IEEE Journal of Quantum Electronics, 35, p. 1156.

47. Zhao H., Tan Y., Zhang R., Zhao Y.-J., Zhang C.-L., Zhang X.-C. & Zhang L.-L. (2021). Molecular dynamic investigation of ethanol-water mixture by terahertz-induced Kerr effect, Optics Express, 29, pp. 36379–36388.

48. Xiao W., Zhang M.-H., Zhang R., Zhang C.-L. & Zhang L.-L. (2023). Highly efficient coherent detection of terahertz pulses based on ethanol, Applied Physics Letters, 122, p. 061105.
49. Tcypkin A., Zhukova M., Melnik M., Vorontsova I., Kulya M., Putilin S., Kozlov S., Choudhary S. & Boyd R. W. (2021). Giant third-order nonlinear response of liquids at terahertz frequencies, Physical Review Applied, 15, p. 054009.
50. Oh T. I., You Y. S. & Kim K. Y. (2012). Two-dimensional plasma current and optimized terahertz generation in two-color photoionization, Optics Express, 20(18), pp. 19778–19786.
51. Kress M., Löffler T., Eden S., Thomson M. & Roskos H. G. (2004). Terahertz-pulse generation by photoionization of air with laser pulses composed of both fundamental and second-harmonic waves, Optics Letters, 29(10), pp. 1120–1122.

Index

www.ingramcontent.com/pod-product-compliance
Lightning Source LLC
Chambersburg PA
CBHW050545190326

41458CB00007B/1921